Human Factors in Land Use Planning and Urban Design

HUMAN FACTORS AND SOCIO-TECHNICAL SYSTEMS

PUBLISHED TITLES

**Human Factors in Land Use Planning and Urban Design:
Methods, Practical Guidance, and Applications**
Nicholas J. Stevens, Paul M. Salmon, Guy H. Walker, Neville A. Stanton

Human Factors in Land Use Planning and Urban Design

Methods, Practical Guidance, and Applications

Nicholas J. Stevens, Paul M. Salmon,
Guy H. Walker, and Neville A. Stanton

CRC Press
Taylor & Francis Group
Boca Raton London New York

CRC Press is an imprint of the
Taylor & Francis Group, an **informa** business

CRC Press
Taylor & Francis Group
6000 Broken Sound Parkway NW, Suite 300
Boca Raton, FL 33487-2742

First issued in paperback 2023

© 2018 by Taylor & Francis Group, LLC
CRC Press is an imprint of Taylor & Francis Group, an Informa business

No claim to original U.S. Government works

ISBN 13: 978-1-03-256988-8 (pbk)
ISBN 13: 978-1-4724-8270-9 (hbk)

DOI: 10.1201/9781315587363

Visit the Taylor & Francis Web site at
http://www.taylorandfrancis.com

and the CRC Press Web site at
http://www.crcpress.com

Contents

V

Preface

I began the journey to publish this human factors (HF) methods book for land use planning and urban design (LUP & UD) in 2012. It started like many things we do, as a consequence of circumstance. At that time, Professor Paul Salmon arrived at the University of the Sunshine Coast, Queensland, Australia, and was located four doors down from my office. We met and began discussing our disciplines—we can admit today about which neither knew much of the other, despite both being significant to our daily lives. It quickly became apparent that there were many similarities between them, not the least of which is the design and evaluation of the interfaces between humans and their environment. We recognized the value in seeking to apply HF methods to the persistent challenges that face LUP & UD.

I would argue that although built environment professionals recognize the importance and impacts of urban development for social, economic, and ecological outcomes, we continue to struggle with managing the complexity of such interdependent systems. In part, because as a discipline we have not assembled a reliable and replicable set of methods to explore the design of complex and multifaceted urban environments, nor evaluate the varied processes and practices we already have. As a consequence, we perpetuate piecemeal solutions and lean on fragmented decision-making as we seek to address often competing priorities. We work independently within the component disciplines of our urban settings—the impacts of architecture, town planning, engineering, landscape architecture, transport planning and psychology, utility providers, community developers, and so on—are rarely considered together within complex urban systems. There has been a tendency from researchers, practitioners, and governments to concentrate on arenas of special interest or political priority therein dealing with many matters in relative isolation.

What is certain is the need for new empirical and scientific approaches to address resource-constrained urban futures. In short, ways to better understand the efficient use of the urban environments we currently have. We need the means to explore the possibilities and implications of value adding existing transport corridors; how can we optimize residual land within current city footprints; how can we make spaces and places that are well used and well loved by all the community; how is it possible to incentivize mixed use, higher density living, supported by adequate utilities and social infrastructure. The *business as usual* approach of continued peripheral urban expansion is neither sustainable from a resource perspective, nor enduring from a community perspective. While this book is not intended as the answer, it does provide the means to better understand the complex systems and subsystems that make up our cities and towns.

The purpose of this book is to offer LUP & UD further tools to rigorously and reliably explore the design and redesign of the environments around us. It provides new ways to interpret urban space and consider context-sensitive analysis of LUP & UD challenges and opportunities. The methods in this book allows for the consideration of the technical aspects of the built environment with the necessary experience; and human-centered approaches to our urban and regional environments.

This book has been constructed so that students, practitioners, and researchers with an interest in one particular area of HF can read the chapters independently from one another. Each category of HF methods is treated as a separate chapter containing an introduction to the area of interest and an overview of the range of applicable methods, including practical guidance on how to apply them. Each of the chapters is therefore self-contained, so those wanting to explore a particular topic can simply choose to read the chapter relevant to their needs.

We feel this book will be of interest to those who are involved in the design and evaluation of land use planning processes and urban spaces. This includes the range of disciplines associated with our built environments, such as architecture, town planning, engineering, landscape architecture, and community development. It will also provide significant guidance for students and researchers within those disciplines seeking to learn more about HF methods. They can apply the methods to their own area of study and use the material within this book to prepare case studies, coursework, and research theses.

The prospect of HF methods being applied in LUP & UD by practitioners and researchers is an exciting one. The testing of theory and methods across domains can only provide new knowledge for all disciplines involved. To this end, it is our hope that this book provides LUP & UD with new approaches to investigate the human condition within the complex systems that are in our cities and towns.

Defining Land Use Planning and Urban Design

Land use planning and urban design (LUP & UD) may mean different things to different people. In the context of this book the term land use planning describes the practice and processes associated with the urban and regional strategic planning. That is, the identification, assessment, and management of the resources and contexts within built and natural environments. Land use planning occurs at a variety of scales from national, to regional, and to local and site-spatial contexts and often deals with the citywide organization of urban form.

Urban design is considered here as the creative delivery of those land use planning objectives at the human and neighborhood scale. Urban design considers the intricacy and optimization of the human experience within our cities and towns. Together, LUP & UD represents an array of important stakeholders and communities, who are responsible for everything from long-term, large-scale strategic visions, right through to the pop-up and tactical urbanism approaches found in the day-to-day vibrancy of urban life.

Nicholas J. Stevens

Acknowledgments

This work would not have been possible without the support of the staff and students within both the Centre for Human Factors and Sociotechnical Systems and the Urban Design and Town Planning Program at the University of the Sunshine Coast, Queensland, Australia. For: ELS, MCS & AIS.

Authors

Nicholas J. Stevens, PhD, is a landscape architect and an urban planner. He is the deputy director of the Centre for Human Factors and Sociotechnical Systems at the University of the Sunshine Coast (USC), Queensland, Australia, where he leads the Land Use Planning and Urban Design Research Theme. He is also the program leader for the Bachelor of Urban Design and Town Planning (Honors) and the Master of Regional and Urban Planning (Research) at USC. His research explores the concept of sociotechnical urbanism and the application of ergonomics and human factors (HF) methods to the perennial problems of land use planning and urban design (LUP & UD). Nicholas has also published widely in the areas of airport and regional land use development and examining the concepts of the Airport City and Aerotropolis. In 2015, Dr. Stevens was awarded an Australian Government Citation for Teaching Excellence for his development of design studio-based curricula. He currently serves on the Planning Institute of Australia's organizing committee and the Urban Design Advisory Panel for the Sunshine Coast, Queensland, Australia.

Paul M. Salmon, PhD, is a professor in Human Factors (HF) and is the creator and Director of the Centre for Human Factors and Sociotechnical Systems at the University of the Sunshine Coast. He currently holds an Australian Research Council Future Fellowship in the area of transportation safety and leads major research programs in the areas of road and rail safety, cybersecurity, and led outdoor recreation. His current research interests include accident prediction and analysis, distributed cognition, systems thinking in transportation safety and healthcare, and HF in elite sports and cybersecurity. Paul has coauthored 11 books, more than 150 journal articles, and numerous conference articles and book chapters. He has received various accolades for his contributions to research and practice, including the UK Ergonomics Society's President's Medal, the Royal Aeronautical Society's Hodgson Prize for best research and paper, and the University of the Sunshine Coast's Vice Chancellor and President's Medal for Research Excellence. In 2016, Paul was awarded the Australian Human Factors and Ergonomics Society Cumming Memorial Medal for his research contribution.

Guy H. Walker, PhD, is an associate professor within the Institute for Infrastructure and Environment at Heriot-Watt University, Edinburgh, Scotland. He lectures on transportation engineering and HF and is the author/coauthor of more than 100 peer-reviewed journal articles and 13 books. He has been awarded the Institute for Ergonomics and Human Factors (IEHF) President's Medal for the practical application of ergonomics theory and Heriot-Watt's Graduate's Prize for inspirational teaching. Dr. Walker has a BSc honors degree in psychology from the University of Southampton, Southampton, United Kingdom; a PhD in human factors from Brunel University, London, United Kingdom; he is a fellow of the Higher Education Academy and a member of the Royal Society of Edinburgh's Young Academy of Scotland. His novel multidisciplinary research has featured in the popular media, from national newspapers, TV, and radio through to an appearance on the Discovery Channel.

Neville A. Stanton, PhD, DSc, is a chartered psychologist, chartered ergonomist, and chartered engineer. He holds the chair in human factors engineering in the Faculty of Engineering and the Environment at the University of Southampton, Southampton, United Kingdom. He has degrees in occupational psychology, applied psychology, and human factors (HF) engineering. His research interests include modeling, predicting, analyzing, and evaluating human performance in systems as well as designing the interfaces and interaction between humans and technology. Professor Stanton has worked on design of automobiles, aircraft, ships, and control rooms over the past 30 years on a variety of automation projects. He has published 40 books and more than 300 journal papers on ergonomics and HF. His work has been recognized through numerous awards, including the Hodgson Prize, awarded to him and his colleagues in 2006 by the Royal Aeronautical Society for research on design-induced, flight-deck, error. He has also received numerous awards from the Institute of Ergonomics and Human Factors in the United Kingdom and in 2014 the University of Southampton awarded him a Doctor of Science for his sustained contribution to the development and validation of HF methods.

1

Introduction to Human Factors Methods in Land Use Planning and Urban Design

1.1 What Is Human Factors?

Human factors (HF) is the discipline dedicated to understanding the interactions and interface of humans with their environments. Through the application of theory, principles, and methods, HF practitioners seek to optimize human behavior, well-being, and system performance. The discipline contributes to the design and evaluation of systems that are compatible with the needs, limitations, and abilities of people. Most often associated with systems of work, it is a multidisciplinary field that bridges psychology, industrial and systems engineering, and computer, safety, and sports sciences. It provides an end-user focused approach for better understanding measured behavior and capabilities, rather than assumptions or trial-and-error, within complex systems of all kinds.

1.2 What Is a Human Factors Problem?

Most readers will be able to identify an example of an HF problem from their own experience of work, study, or just their daily lives. An HF problem will more than likely possess some, or all, of the following attributes. It will be a problem that impacts negatively on individual behavior and overall system performance. It will involve humans in systems who are not behaving as they were expected to because elements of the system were not designed to fit their needs and capabilities. These elements may include the artifacts they are using, the physical environment, the training they have received, the procedures they are working too, the other humans they are interacting with, and so on. It will be a problem that existing methods of design, evaluation, and procurement have somehow not captured, despite in-depth testing and analysis. Above all, it will usually be frustratingly resistant to a whole range of purely technical interventions.

These HF problems impact our daily lives. Their impacts range from minor frustrations, such as cumbersome and difficult to use products, to major catastrophes with significant injuries, fatalities, and social and economic costs. The focus of the HF discipline is to remove these issues through informed system design that is based on an understanding of human and system behavior and the factors that influence it.

The impetus for this book emerged from the realization that HF can play a key role in the analysis and design of our urban environment. Indeed, the description of HF problems above certainly rings true when considering the challenges faced by the disciplines working in land use planning and urban design (LUP & UD). The authors recognize that there are significant parallels between the HF and LUP & UD disciplines. In fact, it is many of these parallels that make this suite of HF methods relevant and effective for exploring our cities and regions. First and foremost of these is that both disciplines recognize that human behavior almost always occurs within systems that are complex in nature.

1.3 Cities as Complex Systems

A complex system, in its simplest sense, is a system with a large number of elements that exchange stimuli with each other and with their environment (Batty, 2007; Ottino, 2003). As such any complex system, including cities, will display specific properties. For demonstrating city complexity, Batty (2007) refers to Durlauf (2005) who states these properties are *nonergodicity*, *phase transition*, *emergence*, and *universality*. *Nonergodicity* is defined as a system that lacks probable behavior over time and, in the context of cities, can be characterized by exogenous shocks—for example, economic, environmental, social (Durlauf, 2005). *Phase transition* refers to a complex system having *tipping points* in which a convergence of elements can change the system. In recognizing the first two properties, *emergence* refers to the new systems properties that arise from the interaction of system components. It is the evolution of the system and is true to the notion that a *system is greater than the sum of its parts* (Batty, 2007). It is emergent changes that, from a positive perspective, can lead to innovation, novelty, and surprise, whereas from an HF safety and risk management approach may represent adverse events or *accidents* within the system (Rasmussen, 1997). Finally, *universality* refers to the system property that when examined at different times and spatial scales, the system is able to be recognized as the same (Durlauf, 2005).

Our cities and their urban and regional environments display all these properties. In fact, cities and urban systems have long been recognized as complex. As Batty (2007) highlights, general system theory provided the early impetus (1950s) as an attractive description of cities, emerging then into the top–down *city as machine* (e.g., Corbusier, 1967, engineered systems) to the more bottom–up *city as organism* (e.g., Holling and Goldberg, 1971, biological systems). As machine, cities were conceived as systems and subsystems which could be influenced by system control, and therein they could be better understood and even kept on task. Early work on supply and demand influences on transport, and land use integration reflects this logic (Batty, 2007). As organism, McLoughlin's (1969) *Urban and Regional Planning: A Systems Approach* sought to provide a framework for the "emerging problems of understanding and planning of cities and towns" (p. 16). It was an approach that conceived the complex systems of human activity in the *whole context of the planets ecological systems*. Whilst not the only theorist to conceive cities in this way, McLoughlin's work was critiqued at the time as an oversimplification of the processes by which decisions are made (Faludi, 1973).

From these foundational explorations of planning and systems theory, through to the present, it is arguably Professor Michael Batty who has the most articulate approach to the idea of our cities as complex systems (Batty, 1971, 1976, 2005, 2008, 2009, 2013, 2015). His work focuses on agent-based computer modeling of cities, their visualization, and related spatial analytic methods. The *New Science of Cities* (2013) provides the background, application, and future of how city design and decision-making can be supported by mathematical modeling and simulation. It provides the means to conceive and simulate the necessary bottom–up approaches that can deal with dynamic and unpredictable city systems. It is the evolution, rather than revolution approach to change in city systems, and talks to issues of emergence, which are also fundamental for HF understandings on urban complexity and systems.

Acknowledging previous systems explorations, the work presented in this book endeavors to provide practical and accessible means for practitioners, researchers, and students to empirically examine our complex city systems and to engage in design process that can cope with this complexity.

1.4 Human Factors and Land Use Planning and Urban Design

What is agreed about complex systems is that they cannot be understood by studying the parts in isolation (Batty, 2007; Ottino, 2003). However, without appropriate methods, this has largely been the approach of the LUP & UD disciplines. The use of interdisciplinary methodologies, such as those offered here, presents an alternative to the continued perpetuation of the *predict and provide* mindset upon which we have relied. They perhaps offer more than another set of normative principles or descriptive visions of technology-rich futures. The *business as usual* approach to urban and regional development is no longer sufficient.

Urbanization and development are occurring at a quickening pace on a global scale. In 1800, 2% of the population lived in urban centers; today, it is 54%; in 2050, it will be closer to 65% (UN, 2014). On World Health Organization's projections, it will be 6.5 billion people living in cities worldwide. Right now, either through childbirth or migration, there is a net urban movement of more than 1 million people to cities every week. New ways to explore the use and reuse of our cities are needed—from the individual site to the strategic nation-building initiatives. Ways in which practitioners, politicians, researchers, and the community can conceive the complex and competing demands of our urban and regional environments.

The current book is not the panacea, but a shift to the HF systems thinking philosophy has important implications for our ability to explore and design for complex environments. Significantly and most importantly it recognizes that the overall system itself is taken as the unit of analysis and must be studied in the context of wider organizational, social, and political factors. For example, although individual physical and cognitive processes should be examined, the systemic factors influencing them should also be considered. This approach to understanding behavior and optimizing system design is now widespread in most safety critical domains including surface transportation (Read et al., 2017; Salmon et al., 2016a), aviation (Stanton et al., 2016), maritime (Lee et al., 2017), defense (Stanton et al., 2010), mining (Donovan et al., 2017), and it is time for LUP & UD to explore the possibilities.

This book focuses on the methodological legacy and accomplishments of the HF discipline and uses them as a launching pad to generate new knowledge for LUP & UD. As such, the work, tasks, and outputs of LUP & UD must correctly be viewed as complex systems. What LUP & UD disciplines may refer to colloquially as a project, design, setting, or environment is considered in the context of this book to represent a *system* or indeed a subsystem.

The authors recognize important parallels between HF and LUP & UD. Both disciplines operate within complex systems settings, yet further than that it is possible to recognize these as important *sociotechnical systems* (STS). STS comprise social and technical elements coengaged in the pursuit of shared goals. The interaction of these social and technical aspects creates emergent properties and the conditions for either successful or unsuccessful system performance (Walker et al., 2010). Stripped back, our urban environments comprise people and communities interacting with technology (objects and artifacts) within environments or indeed a range of urban contexts (Stevens, 2016). The purpose and priority of STS is the optimization of people, technology, environments, and researchers in HF identify that STS approaches have some key features. Importantly, they consider safety as an emergent property and recognize that systems and component performance is variable and that systems are often hierarchical structures (Read et al., 2013; Salmon et al., 2010; Stanton et al., 2013a). These considerations are immediately applicable in the priorities for LUP & UD. That is an understanding that city performance is variable and dynamic, safety is a priority, and that cities are systems within which the hierarchy of context from site to strategic is imperative.

Further discipline alignment is reflected, in that neither field resides exclusively within the purview of engineering, nor are they the exclusive domain of social scientists. Both HF and LUP & UD require careful consideration of exacting tolerances while remaining vigilant and inclusive of the human experience within the system.

Both disciplines also share many methods for data collection, including case studies, surveys, interviews, simulations, visualization, and observation. The naturalistic nature of many of the HF methods is fit for purpose in LUP & UD. Significantly, the use of HF methods presents novel insights and applications to the types of qualitative and quantitative data that LUP & UD are already familiar with collecting and collating.

A key differences is that HF have developed a range of methods that allow them to better understand complexity in terms of how components in a system interact to create emergent behaviors. They have established means to include and explore the physical and cognitive processes associated with human and environmental interaction. LUP & UD continues to struggle with this complexity and is where the use of HF methods can assist.

1.5 Human Factors Methods

HF research and practice is underpinned by a suite of ergonomics methods that support the design or analysis of work and tools in relation to individuals, teams, organizations, and even entire systems. These HF methods are used to describe, represent, and evaluate human activity within complex STS. These methods focus on human interactions with other humans, products, devices, or systems and cover a variety of issues ranging from the physical and cognitive aspects of task performance, errors, decision-making, situation awareness, usability, and physical and mental workloads. They are applied by researchers and practitioners for various reasons including to inform system and product design and redesign, to evaluate existing systems, devices, procedures, and programs; for performance evaluation; for theoretical development; and for training and procedure design.

For the purpose of this book, the HF methods available can be categorized as follows:

1. *Data collection methods.* The starting point in any HF analysis, be it system design or evaluation for theoretical development, involves describing existing analogous systems via the application of data collection methods (Diaper and Stanton, 2004). These methods are used to gather specific data regarding a task, device, system, or scenario, and the data obtained are used as the input for the HF analyses methods described in the following.

2. *Task analysis methods.* Task analysis methods (Annett and Stanton, 2000) are used to describe task and systems and typically involve describing activity in terms of the goals and physical and cognitive steps required. Task analysis methods focus on *what an operator* (the human in the system) is required to do, in terms of actions and/or cognitive processes to achieve a system goal. In recent times, there has been increasing emphasis on the use of task analysis methods to go beyond the operator and describe *what a system* is required to do (Salmon et al., 2016).

3. *Cognitive task analysis (CTA) methods.* CTA methods (Schraagen et al., 2000) focus on the cognitive aspects of task performance and are used for "identifying the cognitive skills, or mental demands, needed to perform a task proficiently" (Militello and Hutton, 2000, p. 90) and describing the knowledge, thought processes, and goal structures underlying task performance (Schraagen et al., 2000). CTA method outputs are used for a variety of different purposes, including, to understand in-depth decision-making processes and the factors influencing them, to inform the design of new technology, systems, procedures, and processes, for the development of training procedures and interventions, for allocation of functions analysis, and for the evaluation of individual and team performance within complex STS.

4. *Error identification methods.* Although there is now less emphasis on human error as it is seen as a consequence rather than a cause of incidents, error identification remains a key concept in accident analysis and prevention. Error identification methods (Kirwan, 1992) use taxonomies of (human) error modes and performance-shaping factors to identify any errors that might occur during a particular task. They are based on the premise that, provided one has an understanding of the task being performed and the technology being used, one can identify the errors that are likely to arise.

5. *Accident analysis methods.* Although both error identification and error analysis methods can be used to analyze accidents, a subsection of methods designed *specifically* to focus on accident analysis is established in HF. Accident analysis methods are employed to derive an accident etiology and to identify contributory factors in the deviation from safe performance. Salmon et al. (2010b) identified over 30 accident analysis-related methods, illustrating the prominence of accident analysis methods in contemporary HF research and practice.

6. *Situation awareness assessment.* Situation awareness refers to an individual's, team's, or system's awareness of *what is going on* during task performance. Situation awareness measures (Salmon et al., 2009) are used to measure and/or model situation awareness during task performance. Such analyses are used to understand the factors that limit or degrade situation awareness or to test the impact of new technologies, procedures, and training systems on situation awareness.

7. *Mental workload assessment methods.* Mental workload represents the proportion of cognitive resources that are demanded by a task or series of tasks. Mental workload measures are used to determine the level of operator mental workload incurred during task performance. Similar to situation awareness measures, they are most often used to test the impacts of new technologies, procedures, and training systems on user mental workload.

8. *Interface evaluation methods*. A poorly designed interface can lead to unusable products, user frustration, user errors, inadequate performance, and increased performance times. Interface evaluation approaches (Stanton and Young, 1999) aim to improve interface design by understanding or predicting user interaction with the product, service or environment in question.

9. *System analysis methods*. System analysis methods can be used to provide exhaustive analyses of complex STS and their behavior. In recent times, there has been shift in the focus from studying individual and team behavior to the study of the behavior of overall systems (Salmon et al., 2017b). This so-called systems-thinking approach involves taking the overall system as the unit of analysis, looking beyond individuals, and considering the interactions between humans and between humans and artifacts within a system. This view also encompasses factors within the broader organizational, social, or political system in which processes or operations take place.

1.5.1 When to Apply the Methods

A key strength of HF methods is that they can be applied across a system design lifecycle. Indeed, the benefits of HF are only realized when HF knowledge and methods are applied from design concept stage through to the fully operational system stage. Similar to LUP & UD, HF input into projects is best considered in the early phases of design. Both disciplines recognize that considerable time, effort, and expense can often be saved by early intervention rather than being faced with a completed operational system (or project) that requires considerable effort to redesign. Unfortunately, this is a common problem.

Fortunately, many HF methods are flexible with regard to the design stage they could be applied to, even if the system itself is no longer as flexible in terms of subsequent changes. There are methods in this book, which lend themselves well to being applied at the very early stages of design. There are also methods explained which may be used in a predictive as well as an evaluative manner. Called *analytical*

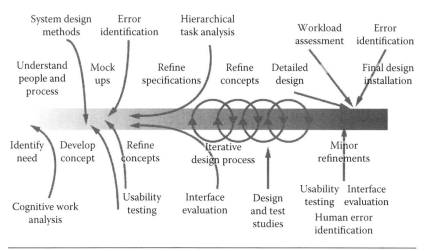

Figure 1.1   Application of HF methods by phase of the design process. (Adapted from Stanton, N. A. et al., *Human Factors Methods: A Practical Guide for Engineering and Design*, 2nd ed, Ashgate, Aldershot, UK, 2013.)

prototyping, this is the process of applying HF insights into systems that do not yet exist in physical form and these approaches offer significant insight for LUP & UD. Figure 1.1 presents a generic design process in which different HF methods become applicable and useful at different stages.

At the start, we begin with methods that are suited to *analytical prototyping* and to modeling the constraints of a particular problem domain to reveal opportunities for unexpected behaviors. The analysis would then proceed forward with analyses of (human) error, usability, and interface evaluation, amongst others. Each method would be chosen to suit the particular stage of the design and urban development life cycle. For example, in the early stages, methods would be chosen to enable designers and engineers to diagnose important dimensions of their proposals and the systems. In later stages, methods would be chosen, which reflect the fact that a physical manifestation of the system now exists and that users themselves can start interacting with it.

Given that most HF and LUP & UD problems emerge from unexpected interactions at the boundary between people and systems, the need to engage in an evolutionary, iterative, *design–test–design* process emerges as a consistent theme in projects within which the authors have worked. This book is not about the LUP & UD process per se but suffice to say we contend that problems may be avoided with systems approaches to LUP & UD, and it is to systems approaches that HF methods lend themselves very well.

1.6 Scientist or Practitioner?

The current book seeks to provide a series of methods that are tested within the HF discipline and demonstrate their efficacy in providing richer more rigorous and enquiring approaches to LUP & UD. The intention is not to restrict intuitive processes of design but to provide new ways to explore the necessary efficiencies that must be gained from our existing and next generation-built environments. The planning *as best we can* and *business usual approaches* to the design and function of our cities have run their course. We contend that the HF methods and discipline, who deal with the foibles of human experience, up against engineered have much to offer LUP & UD.

What is important is that the approaches we offer here to LUP & UD are built on a foundation of robust human science, with actionable methods long forming a major part of the HF discipline. The *International Encyclopaedia of Human Factors and Ergonomics* (Karwowski, 2001) has an entire section devoted to methods and techniques. In a recent review of such methods, Stanton et al. (2013) identified well over 150 methods. The importance of HF methods to process and design cannot be overstated. These methods offer the engineer, the designer, the specialized HF, and now LUP & UD practitioners a structured approach to the analysis and evaluation of practical problems. The overall approach we offer LUP & UD can be described using the scientist–practitioner model (Stanton, 2005). As a scientist, the process of applying these methods is as follows:

- Extending the work of others
- Testing theories of human-system performance
- Developing hypotheses
- Questioning everything
- Using rigorous data collection and analysis techniques
- Ensuring the repeatability of results
- Disseminating the findings of studies

As a practitioner, the application of these methods is as follows:

- Addressing real-world problems
- Seeking the best compromise under difficult circumstances
- Looking to offer the most cost-effective solution
- Developing demonstrators and prototype solutions
- Analyzing and evaluating the effects of change
- Developing benchmarks for best practice
- Communicating findings to interested parties

In applying the methods contained in this book, you will work somewhere between the poles of scientist and practitioner, varying the emphasis of your approach depending upon the problems that you face. HF methods are useful in the scientist–practitioner model for LUP & UD because of the structure and potential for repeatability that they offer. There is an implicit guarantee in the use of methods that, provided they are used properly, they will produce certain types of useful products. HF methods are a route to making systems design and outputs more accessible to all (Diaper, 1989; Stanton and Young, 2003). This is critical for LUP & UD given the multidisciplinary, human-centered, engineering, and design-based nature of the opportunities and challenges faced.

1.7 Reliability and Validity

To the engineer or designer, the human sciences in general (and possibly HF and LUP & UD in particular) may fall victim to the popular, albeit wholly inaccurate, perception of being rather *woolly* fields. This is not so. Although LUP & UD has lacked a comprehensive set of methods, HF has not. Both disciplines deal with problems that may seem alarmingly loose in definition and that do not conform to any readily identifiable chain of cause and effect. However, in facing such problems, HF methods provide a welcome source of structure and rigor. Stanton (2016) in particular has pioneered studies on the reliability and validity of HF methods (Stanton and Young, 1999a, b; Stanton et al., 2014). These studies have increased confidence in HF methods and demonstrated how reliability and validity data should be reported. The work has also shown the relative cost-effectiveness of different HF methods and how long training in the methods takes.

Researchers have identified a dichotomy of HF methods: analytical methods and evaluative methods (Annett, 2002). They argue that analytical methods (i.e., those methods that help the analyst gain an understanding of the mechanisms underlying the interaction between human and their environments) require construct validity, whereas evaluative methods (i.e., those methods that estimate parameters of selected interactions between human and their environments) require predictive validity. Construct and criterion referenced validity play a role in the development of HF theory itself. There is a difference between construct validity (how acceptable the underlying theory is), predictive validity (the usefulness and efficiency of the approach in predicting the behavior of an existing or future system), and reliability (the repeatability of the results). This distinction is made in Table 1.1.

This presents an interesting question. Are the methods really mutually exclusive? Some HF methods appear to have dual roles which implies that they must satisfy both criteria. It is plausible, however, as Baber (2005) argues in terms of evaluation, that the approach taken will influence which of the purposes one might wish to emphasize. The implication is that the way in which one approaches a problem— or, in other words, where on the scientist–practitioner continuum one places oneself—could well have a bearing on how a method is employed. At first glance (particularly from a *scientist* perspective), such a *pragmatic* approach appears highly dubious. If we are selecting methods piecemeal to satisfy contextual requirements, how can we be certain that we are producing useful, valid, reliable output? Although it may be possible for a method to satisfy three types of

Table 1.1   Annett's Dichotomy of Ergonomics Methods

	ANALYTIC	EVALUATE
Primary purpose	Understand a system	Measure a parameter
Examples	Task analysis, training needs analysis, and so on	Measures of workload, usability, comfort, fatigue, and so on
Construct validity	Based on an acceptable model of the system and how it performs	Is consistent with theory and other measures of parameters
Predictive validity	Provides answers to questions, for example, structure of tasks	Predicts performance
Reliability	Data collection conforms to an underlying model	Results from independent samples agree

Source: Annett, J., *Theoretical Issues in Ergonomics Science*, 3(2), 229–232, 2002.

validity—construct (i.e., theoretical validity), content (i.e., face validity), and predictive (i.e., criterion-referenced empirical validity), it is
not always clear whether this arises from the method itself or from the
manner in which it is applied. The solution, simply stated, is that care
needs to be taken before embarking on any application of methods to
make sure that one is attempting to use the method in the spirit in
which it was designed.

1.8 Which Method to Use?

How do you decide which of the 30 HF methods contained in this
book to apply to a particular LUP & UD problem? Some urban
development challenges require only a basic level of HF insight and
a correspondingly basic methodological intervention. Other problems
require greater levels of sophistication, and determining an appropriate set of methods (because individual methods are rarely used alone
in such cases) requires some planning and preparation. Increasingly
complex systems require you to have a flexible strategy, so pilot studies
are often helpful in scoping out the problem before a detailed study is
undertaken. From a practitioner perspective, the time taken to carry
out pilot studies might simply be unavailable. However, we would
argue that there is no harm in running through a selection of methods
as a form of *thought-experiment* to ascertain what type of output each
method is likely to produce, and deciding whether or not to include
a method in the battery of methods that will be applied. Although it
is important not to rely too heavily on a single approach, there is no
guarantee that simply throwing a lot of methods at a problem will
guarantee useful results. An informed approach is needed.

Faced with fact that LUP & UD issues are often particularly
complex and multidimensional, a fundamental question to ask is
"What is the nature of the problem that my selection of methods is
aiming to resolve?" The notion of an *HF problem space* could serve as a
useful device in shaping your thinking regarding the choice of methods for LUP & UD. If your particular problem can be defined as having low levels of change over time, a small number of interconnected
parts and the principles of the system's operation are well understood,
then methods suited to this more *deterministic* type of problem may be
appropriate. In other words the use of methods which break a problem

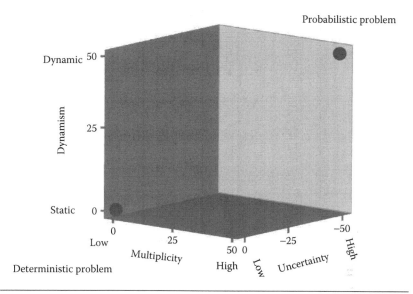

Figure 1.2 The systems design problem space is a device that can be used to shape thinking as to what method might be appropriate to what problem. (From Stanton, N. A. et al., *Human Factors Methods: A Practical Guide for Engineering and Design*, 2nd ed, Ashgate, Aldershot, UK, 2013.)

down into component parts on the tacit assumption that the whole can be no more, or no less, than the sum of those parts. At the other end of the spectrum are complex problems, those with multiple interconnected parts, high rates of change, and high rates of uncertainty. Faced with problems similar to these, those that are more *probabilistic* in nature might prompt you to use methods more closely aligned to systems thinking and of a *formative* nature (focusing on what *could* happen rather than what *should* happen). Figure 1.2 presents the systems design problem space.

1.9 Using the Book

The current book includes the phrase *a practical guide* in its title for a reason. It does not dwell extensively on theory; rather, it focuses on the more pragmatic question of what HF methods are, their advantages and disadvantages, and step-by-step guidance on how to carry them out yourself in the context of LUP & UD. In putting this book together, we reviewed a large collection of contemporary HF methods over two stages. First, a review of existing HF methods and techniques was conducted to identify those with the greatest applicability to LUP & UD. Second, a screening process was employed to remove any methods that require more than paper and pencil to conduct. The reason for this latter criterion was not to disparage any of the computer-based tools on the market but to focus on those techniques that the practitioner could use without recourse to specialized equipment or providers. Third, the methods selected for review were then described and evaluated using a set of predetermined criteria to give you confidence in the breadth and depth of the methods selected for inclusion.

The criteria were designed not only to establish which of the techniques was the most suitable for use in the design and evaluation of LUP & UD systems but also to provide a standardized, simplified way

Table 1.2 Descriptions of Method Review Criteria

CRITERIA	DESCRIPTION OF CRITERIA
Name and acronym	The name of the technique or method and its associated acronym
Background and applications	This section introduces the method, its origins and development, the domain of application of the method, and also application areas in which it has been used
Domain of application	This describes the domain that the technique was originally developed for and applied in
Application in land use planning and urban design	This describes the applications and potential for the method for use by land use planning and urban design practitioners, researchers, and students
Procedure and advice	This section describes the procedure for applying the method as well as general points of expert advice
Flowchart	A flowchart is provided, depicting the method's procedure
Advantages	Lists the advantages associated with using the method in the design of systems
Disadvantages	Lists the disadvantages associated with using the method in the design of systems
Example	An example (or examples) of the application of the method is provided to show the method's output
Related methods	Any closely-related methods are listed, including contributory and similar methods
Approximate training and application times	Estimates of the training and application times are provided to give the reader an idea of the commitment required when using the technique
Reliability and validity	Any evidence on the reliability or validity of the method is cited
Tools needed	This describes any additional tools required when using the method
Recommended texts	Lists recommended further reading on the method and the surrounding topic area

of communicating *how to* perform the method in practice. Therefore, the output is designed to act as a manual, with the headings in Table 1.2 serving as the structure for each of the methods contained in this book.

The current book has been designed for you to consult for advice and guidance on which methods have potential application to your LUP & UD problem and how to actually use any given method in practice. The book is also designed to help you understand the interdependencies between methods and which method outputs are required to act as inputs for other methods. HF methods can enable LUP & UD researchers and practitioners to add considerable value to current and future city and urban development systems. This book presents an actionable set of methods that can be put to immediate use in achieving this aim.

2

DATA COLLECTION METHODS

2.1 Introduction

The starting point of any analysis will be the scoping and definition of expected outcomes; for example, this might mean defining hypotheses or determining which questions the analysis is intended to answer. Following this stage, effort normally involves collecting specific data regarding the land use planning and urban design (LUP & UD) system, activity, and people that the analysis effort is focused upon. In the design of novel or new systems, information regarding activity in similar, existing systems is required. This allows the planning and design team to evaluate existing or similar systems in order to determine existing design flaws and problems and also to highlight efficient aspects that may be carried forward into the new design. The question of what constitutes a *similar* LUP & UD system is worth considering at this juncture. If we concentrate solely on the current generation of systems (with a view to planning the next generation), then it is likely that any design proposals would simply be modifications to current technology, approaches, or practice. Although this might be appropriate in many instances, it does not easily support original design and redesign for city systems (which might require a break with current thinking). An alternative approach is to find systems that reflect some core aspect of current work and then attempt to analyze the activity within these systems. Thus, in designing novel technology to support newspaper editing, production, and layout planning, Bødker (1988) focused on manual versions of the activities rather than on the contemporary word processing or desktop-publishing systems. An obvious reason for doing this is that the technology (particularly at the time of her study) would heavily constrain the activity that people could perform, and these constraints might be appropriate for the limitations of the technology but not supportive of the goals and activities of the people working within the system. Therefore, when considering the design of LUP & UD systems, it is important to consider systems that reflect the outcomes or indeed efficiencies that are desired for the new system. In a similar way, water-sensitive urban design (Wong, 2006) looks to natural systems to deal with the challenges of urban storm water. Thus, it can be highly beneficial to look at activity away from the *current* technology for several reasons as follows:

1. Avoiding the problems of technology constraining possible activity
2. Allowing appreciation of the fundamental issues relating to the goals of the system (as opposed to understanding the manner in which particular technology needs to be used)
3. Allowing (often) rapid appreciation of basic needs without the need to fully understand complex technology

The human factors (HF) evaluation of existing, operational city systems and design (e.g., usability, error analysis, and task analysis) also requires that specific data regarding task performance in the system under analysis are collected, represented, and analyzed accordingly. This is something that has been inherently lacking within the consideration of current city systems and design and is a key to their appropriate redesign. Data collection methods therefore need to represent the cornerstone of any analysis effort. Such methods are used by the LUP & UD practitioner to collect specific information regarding the system, activity, or artifact under analysis, including the nature of the activity conducted within the system, the individuals performing the activity, the component task steps and their sequence, the technological artifacts used by the system, and the people necessary in performing the tasks (controls, displays, communication technology, etc.), the system environment, and also the organizational and jurisdictional environment within which the system exists.

The importance of an accurate representation of the system or activity under analysis cannot be underestimated and is a necessary prerequisite for any further analysis efforts. As we noted earlier, the starting point for designing future city systems is a description of the current or analogous system, and any inaccuracies within the description could potentially hinder the design effort. Data collection methods are used to collect the relevant information that is used to provide this description of the system or activity under analysis. There are a number of different data collection methods used and available to the LUP & UD practitioner, including observation, interviews, questionnaires, analysis of artifacts, design and sites, usability metrics, and the analysis of performance. Often, data collected through the use of these methods can be utilized as the starting point or input for a range of HF methods.

The main advantage associated with the application of data collection methods is the high volume and utility of the data that are collected. The analyst(s) using the methods also have a high degree of control over the data collection process and are able to direct the data collection procedure as they see fit. Despite the usefulness of data collection methods, there are a number of potential problems associated with their use. For example, one problem associated with the utilization of data collection methods such as interviews, observational study, and questionnaires is the high level of resource usage incurred, particularly during the design of data collection procedures. The design of interviews and questionnaires is a lengthy process, involving numerous pilot runs and reiterations. In addition to this, large amounts of data are typically collected, and lengthy data-analysis procedures are common. In addition to the high resource usage incurred, data collection techniques also require access to the system and personnel under analysis, which is often very difficult and time-consuming to obtain. If the data need to be collected during operational design and planning scenarios, getting the required personnel to take part in interviews is also difficult, and questionnaires often have very low return rates, that is, typically 10 percent for a postal questionnaire. Similarly, there are difficulties associated with the observation and recording of the public and private users of city systems. A brief description of each of the data collection methods is given in Sections 2.1.1 through 2.1.3.

2.1.1 Interviews

Interviews offer a flexible approach to data collection and have consequently been applied for a plethora of different purposes. They can be used to collect a wide variety of data, ranging from user perceptions and reactions to space and place, to day-to-day work functions. There are three types of interview available to the practitioner: structured, semistructured and unstructured or open interviews. Typically, participants are interviewed on a one-to-one basis, and the interviewer uses predetermined probe questions to elicit the required information.

2.1.2 Questionnaires

Questionnaires offer a very flexible means of quickly collecting large amounts of data from large participant populations. They have been used in many forms to collect data regarding numerous issues within HF and LUP & UD design and evaluation and can be used to collect information regarding almost anything at all, including usability, user satisfaction, opinions, and attitudes. More specifically, they can be employed throughout the design process to evaluate design concepts and prototypes, to probe user perceptions and reactions, and to evaluate existing systems.

2.1.3 Observation

Observation (and observational studies) is used to gather data regarding activity conducted in complex, dynamic systems. In its simplest form, it involves observing an individual or group of individuals performing work or indeed day-to-day activities. A number of different types of observational study exist, such as direct observation, covert observation, and participant observation. Observation is attractive due to the volume and utility of the data collected and also due to the fact that the data are collected in an operational context. Although, at first glance, simply observing an someone at work, or during the course of their day, seems to be a very simple technique to employ, it is evident that this is not the case, and that careful planning and execution are required (Stanton et al., 2004). Observational techniques also require the provision of technology, such as video and audio-recording equipment. The output from an observational analysis is used as the primary input for many HF techniques, such as task analysis, error analysis, and charting techniques.

2.2 Interviews

2.2.1 Background and Applications

Interviews provide the LUP & UD practitioner and researcher with a flexible means of gathering large amounts of specific information regarding a particular subject. Due to the flexible nature of interviews, they have been used extensively in HF to gather information on a plethora of topics, including system usability, user perceptions, reactions and attitudes, job analysis, cognitive task analysis, error, and many more. Moreover, designing their own interviews, HF practitioners also have a number of specifically designed interview techniques at their disposal. For example, the critical decision method (Klein and Armstrong, 2004) is a cognitive task analysis technique that provides the practitioner with a set of cognitive probes designed to elicit information regarding decision-making during a particular scenario (Chapter 4). The three generic interview *types* typically employed by the HF and LUP & UD practitioner are outlined in the following:

- *Structured*: In a structured interview, the interviewer probes the participant using a set of predefined questions designed to elicit specific information regarding the subject under analysis. The content of the interview (questions and their order) is predetermined, and no scope for further discussion is permitted. Due to their rigid nature, structured interviews are the least popular type of interview. They are only used when the type of data required is rigidly defined, and no additional data are required.
- *Semistructured*: When using a semistructured interview, some of the questions and their order are predetermined. However, semistructured interviews are flexible in that the interviewer can direct the focus of the interview and also use further questions that were not originally part of the planned interview structure. As a result, information surrounding new or unexpected issues is often uncovered during semistructured interviews. Due to this flexibility, the semistructured interview is the most commonly applied type of interview.

- *Unstructured*: When using an unstructured interview, there is no predefined structure or questions, and the interviewer goes into the interview *blind* so to speak. This allows the interviewer to explore, on an ad hoc basis, different aspects of the subject under analysis. Although their flexibility is attractive, unstructured interviews are infrequently used, as their lack of structure may result in crucial information being neglected or ignored.

2.2.2 Focus Group

Although many interviews concentrate on the one-to-one elicitation of information, group discussions can provide an efficient means of canvassing consensus opinion from several people. Ideally, the focus group would contain around five people with similar backgrounds, and the discussion would be managed at a fairly high level, that is, rather than asking specific questions, the analyst would introduce topics and would facilitate their discussion. A useful text for exploring focus groups is Langford and McDonagh (2002).

2.2.3 Question Types

An interview involves the use of questions or probes designed to elicit information regarding the subject under analysis. An interviewer typically employs three different types of question during the interview process. These are closed questions, open-ended questions, and probing questions. A brief description of each interview question type is presented in the following:

- *Closed*: Closed questions are used to gather specific information and typically permit "yes" or "no" answers. An example of a closed question would be "Do you think that X is usable?" The question is designed to gather a "yes" or "no" response, and the interviewee does not elaborate on their chosen answer.
- *Open-ended*: An open-ended question is used to elicit more than the simple "yes"/"no" information that a closed question gathers. It allows the interviewees to answer in whatever way they wish and also to elaborate on their answer. For example, an open-ended question approach to the topic of the usability

of a park shelter would be something like "What do you think about the usability of this park shelter?" By allowing the interviewee to elaborate upon answers given, open-ended questions typically gather more pertinent data than closed questions. However, open-ended question data require more time to analyze than closed question data, and so closed questions are often used.

- *Probing*: A probing question is normally used after an open-ended or closed question to gather more specific data regarding the interviewee's previous answer. Typical examples of a probing question would be "Why did you think that seating was not usable?" or "How did it make you feel when you were using the seating?"

Stanton and Young (1999) recommend that interviewers should begin with a specific topic and probe it further until the topic is exhausted; then they move on to a new topic. They advocate that the interviewer should begin by focusing on a particular topic with an open-ended question, and then, once the interviewee has answered, use a probing question to gather further information. A closed question should then be used to gather specific information regarding the topic. This cycle of open-ended, probe, and closed questions should be maintained throughout the interview.

2.2.4 Domain of Application

Generic.

2.2.5 Application in Land Use Planning and Urban Design

Interviews are one of the most commonly used data collection techniques within LUP & UD.

2.2.6 Procedure and Advice (Semistructured Interview)

There are no set rules to adhere to during the construction and conduction of an interview. The following procedure is intended to act as a set of flexible guidelines for the LUP & UD practitioner.

Step 1: Define the Interview Objective

Initially, before the design of the interview begins, the analyst should clearly define what the aims and objectives of the interview are. Without a clearly defined objective, the focus of the interview is unclear, and the data gathered during the interview may lack specific content. For example, when interviewing an urban designer for a study into the use of public spaces, the objective of the interview would be to discover which uses the designer had considered or had seen in the past, in which part of the public space and when. A clear definition of the interview objectives ensures that the interview questions used are wholly relevant, and that the data gathered are of optimum use.

Step 2: Question Development

Once the aims and objectives of the interview are clearly defined, development of appropriate interview questions can begin. The questions should be developed upon the basis of the overall objective of the interview. In the urban designer's case, examples of pertinent questions would be "What sort of uses have you designed for in the past in public space?" This would then be followed by a probing question such as "Why do you think these uses are important?" or "Where was this space you were designing?" Once all of the relevant questions are developed, they should be put into some sort of coherent order or sequence. The wording of each question should be very clear and concise, and the use of acronyms or confusing terms should be avoided. An interview transcript or data collection sheet should then be created, containing the interview questions and spaces for demographic information (name, age, sex, occupation, etc.) and interviewee responses.

Step 3: Piloting the Interview

Once the questions have been developed and ordered, the analyst should then perform a pilot or trial run of the interview procedure. This allows any potential problems or discrepancies to be highlighted. Typical pilot interview studies involve submitting the interview to colleagues or even performing a trial interview with real participants. This process is very useful in shaping the interview into its most efficient form and allows any potential problems in the data collection procedure to be highlighted and eradicated. The analyst is also given an indication of the type of data that the interview may gather and can change the interview content if appropriate.

Step 4: Redesign Interview Based upon Pilot Run

Once the pilot run of the interview is complete, any changes highlighted should be made. This might include the removal of redundant questions, the rewording of existing questions, or the addition of new questions.

Step 5: Select Appropriate Participants

Once the interview has been thoroughly tested and is ready for use, the appropriate participants should be selected. Normally, a representative sample from the population of interest is used. For example, in an analysis of designing public space, the participant sample would comprise urban designers with varying levels of experience.

Step 6: Conduct and Record the Interview

According to Stanton and Young (1999), the interviewee should use a cycle of open-ended, probe, and closed questions. They should persist with one particular topic until it is exhausted and should then move on to a new one. General guidelines for conducting an interview prescribe that the interviewer is confident and familiar with the topic in question, communicates clearly, and establishes a good rapport with the interviewee. The interviewer should avoid being overbearing and should not mislead, belittle, embarrass, or insult the interviewee. The use of technical jargon or acronyms should also be avoided. It is recommended that the interview be recorded using either audio or visual-recording equipment.

Step 7: Transcribe the Data

Once the interview is completed, the data should be transcribed. This involves replaying the initial recording of the interview and transcribing fully everything that is said during the interview, both by the interviewer and the interviewee.

This is typically a lengthy and laborious process and requires much patience on behalf of the analyst involved. It might be worth considering paying someone to produce a word-processed transcription of the data.

Step 8: Data Gathering

Once the transcript of the interview is complete, the analyst should examine the interview transcript, looking for the specific data that were required by the objective of the interview. This is known as the *expected* data. Once all of the *expected data* are gathered, the analyst should reexamine the interview to gather any *unexpected data*, that is, any extra data (not initially outlined in the objectives) that are unearthed.

Step 9: Data Analysis

Finally, the analysts should then examine the data using appropriate statistical tests, graphs, and so on. The form of analysis used is dependent upon the aims of the analysis but typically involves converting the words collected during the interview into numerical form in readiness for statistical analysis. A good interview will always involve planning, so that the data are collected with a clear understanding of how subsequent analysis will be performed. In other words, it is not sufficient to have piles of handwritten notes following many hours of interviewing and then have no idea what to do with them. A good starting point is to take the transcribed information and then perform some *content analysis*, that is, divide the transcription into specific concepts. Then it can be determined whether the data collected from the interviews can be reduced to some numerical form, for example, counting the frequency with which certain concepts are mentioned by different individuals or the frequency with which concepts occur together.

Alternatively, the content of the interview material might not be amenable to reduction to numerical form, and so it is not possible or sensible to consider statistical analysis. In this case, it is a common practice to work through the interview material and look for common themes and issues. These can be separated out and (if possible) presented back to the interviewees, using their own words. This can provide quite a powerful means of presenting opinion or understanding. If the interview has been video recorded, then it can be useful to edit the video down in a similar manner, that is, to select specific themes and use the video of the interviewees to present and support these themes.

2.2.7 Advantages

- Interviews can be used to gather data regarding a wide range of subjects.
- Interviews offer a very flexible way of gathering large amounts of data.
- The data gathered are potentially very powerful.
- The interviewer has full control over the interview and can direct the interview in any way.
- Response data can be treated statistically.
- A structured interview offers consistency and thoroughness (Stanton and Young, 1999).
- Interviews have been used extensively in the past for a number of different types of analysis.
- Specific, structured HF interview techniques already exist, such as the critical decision method (Klein and Armstrong, 2004).

2.2.8 Disadvantages

- The construction and data-analysis process ensure that the interview technique is a time-consuming one.
- The reliability and validity of the technique are difficult to address.
- Interviews are susceptible to both interviewer and interviewee bias.
- Transcribing the data is a laborious, time-consuming process.
- Conducting an interview correctly is quite difficult and requires great skill on behalf of the interviewer.
- The quality of the data gathered is based entirely upon the skill of the interviewer and the quality of the interviewee.

2.2.9 Approximate Training and Application Times

In a study comparing 12 HF techniques, Stanton and Young (1999) reported that interviews took the longest to train of all the methods due to the fact that the technique is a refined process requiring a clear understanding on the analyst's behalf. In terms of application times, a normal interview could last anything between 10 and 60 min. Kirwan and Ainsworth (1992) recommend that an interview should last a minimum of 20 min and a maximum of 40 min. Although this represents a low application time, the data analysis part of the interview technique can be an extremely lengthy one (e.g., data transcription, data gathering, and data analysis). Transcribing the data is a particularly time-consuming process. For this reason, the application time for interviews is estimated as very high.

2.2.10 Reliability and Validity

Although the reliability and validity of interview techniques are difficult to address, Stanton and Young (1999) reported that in a study comparing 12 HF techniques, a structured interview technique scored poorly in terms of reliability and validity.

2.2.11 Tools Needed

An interview requires a pen and paper and an audio-recording device, such as a digital voice recorder. A PC or Mac with a word-processing package such as Microsoft Word™ is also required to transcribe the data and statistical analysis packages such as SPSS™ may be required for data-analysis procedures.

2.2.12 Flowchart

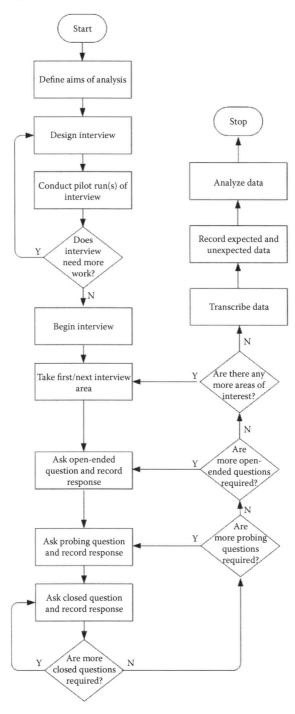

2.2.13 Recommended Text(s)

Jacob, S.A., & Furgerson, S.P., (2012). Writing interview protocols and conducting interviews: Tips for students new to the field of qualitative research. *The Qualitative Report*, *17*(42), 1–10.
Rowley, J., (2012). Conducting research interviews. *Management Research Review*, *35*(3/4), 260–271.

2.3 Questionnaires

2.3.1 Background and Applications

Questionnaires offer a very flexible way of quickly collecting considerable amounts of specific data from a large population sample. They have been used in many forms to collect data regarding numerous issues within both LUP & UD and HF, including usability, user satisfaction, and user opinions and attitudes. More specifically, they can be used in the design process to evaluate concept and prototypical designs to probe user perceptions and to evaluate existing system designs. They can also be employed in the evaluation process to evaluate system usability or attitudes toward a system. Specific questionnaires can be designed and administered upon the basis of the objectives of a particular study. The method description offered here will concentrate on the design of questionnaires, as the procedure used when applying existing HF questionnaire techniques is described in the following chapters.

2.3.2 Domain of Application

Generic.

2.3.3 Application in Land Use Planning and Urban Design

Questionnaires are one of the most commonly used data collection techniques within LUP & UD. They are an approach that most practitioners, researchers, and students will be familiar with.

2.3.4 Procedure and Advice

There are no set rules for the design and administration of questionnaires. The following procedure is intended to act as a set of guidelines to consider when constructing a questionnaire.

Step 1: Define the Study Objectives

The first step involves clearly defining the objectives of the study, that is, what information is wanted from the questionnaire data that are gathered. Before any effort is put into the design of the questions, the objectives of the questionnaire must be clearly defined. It is recommended that the analyst should go further than merely describing the goal of the research. For example, when designing a questionnaire to gather information on the usability of a system or product, the objectives should contain precise descriptions of different usability problems already encountered and descriptions of the usability problems that are expected. In addition, the different tasks involved in the use of the system in question should be defined, and the different personnel should be categorized. What the results are supposed to show and what they could show should also be specified, as well as the types of questions (closed, multiple-choice, open-ended, rating, ranking, etc.) to be used. This stage of questionnaire construction is often neglected, and consequently the data obtained normally reflect this (Wilson and Corlett, 1995).

Step 2: Define the Population

Once the objectives of the study are clearly defined, the analyst should define the sample population, that is, the participants whom the questionnaire will be administered to. Again, the definition of the participant population should be as exhaustive as possible, including defining age groups, different user groups, and different organizations. The sample size should also be determined at this stage. Sample size is dependent upon the scope of the study and also the amount of time and resources available for data analysis.

Step 3: Construct the Questionnaire

A questionnaire typically comprises four parts: an introduction, a participant information section, an information section, and an epilog. The introduction should contain information that lets the participant know who you are, what the purpose of the questionnaire is, and what the results are going to be used for. One must be careful to avoid putting information in the introduction that may bias the participant in any way. For example, describing the purpose of the questionnaire as "determining people's attitudes to expenditure on public art." The classification part of the questionnaire normally contains multiple-choice questions requesting information about the participant, such as age, sex, occupation, and experience. The information part of the questionnaire is the most crucial part, as it contains the questions designed to gather the required information related to the initial objectives. There are numerous categories of questions that can be used in this part of the questionnaire. Which type of question to be used is dependent upon the analysis and the type of data required. Where possible, the type of question used in the information section of the questionnaire should be consistent, that is, if the first few questions are of multiple choices, then all the questions should be kept as multiple choices. The different types of questions available are displayed in Table 2.1. Each question used in the questionnaire should be short in length and worded clearly and concisely, using relevant language. Data analysis should be considered when constructing the questionnaire. For instance, if there is little time available for the data-analysis process, then the use of open-ended questions should be avoided, as they are time-consuming to collate and analyze. If time is limited, then closed questions should be used, as they offer specific data that are quick to collate and analyze. The size of the questionnaire is also important. If it is too large, the participants will not complete the questionnaire, yet a very small questionnaire may seem worthless and could suffer the same fate. The optimum questionnaire length is dependent upon the participant population, but it is generally recommended that questionnaires should be no longer than two pages (Wilson and Corlett, 1995).

Step 4: Piloting the Questionnaire

Wilson and Corlett (1995) recommended that once the questionnaire construction stage is complete, a pilot run of the questionnaire is required. This is a crucial part of the

Table 2.1 Types of Questions Used in Questionnaire Design

TYPE OF QUESTION	EXAMPLE QUESTION	WHEN TO USE
Multiple choice	Approximately how many occasions have you witnessed people using this space? (0–5, 6–10, 11–15, 16–20, more than 20)	When the participant is required to choose a specific response
Rating scales	I found the design unnecessarily complex (strongly agree [5], agree [4], not sure [3], disagree [2], strongly disagree [1])	When subjective data regarding participant opinions are required
Paired associates (bipolar alternatives)	Which of the two tasks "a" + "B" subjected you to the most mental workload? ("a" or "B")	When two alternatives are available to choose from
Ranking	Rank, on a scale of 1 (very poor usability) to 10 (excellent usability), the usability of the design	When a numerical rating is required
Open-ended questions	What do you think of the parks usability?	When data regarding participants' own opinions about a certain subject are required, that is, subjects compose their own answers
Closed questions	Which of the following elements have you used in public open spaces (seats, tables, playground equipment, bathrooms)	When the participant is required to choose a specific response
Filter questions	Have you ever committed an error whilst using the current system interface? ("yes" or "no;" if "yes," go to question 10; if "no," go to question 15)	To determine whether participant has specific knowledge or experience; to guide participant past redundant questions

Source: Stanton, N. A. et al., *Human Factors Methods: A Practical Guide for Engineering and Design*, 2nd ed., Aldershot, Ashgate, 2013a.

questionnaire design process, yet it is often neglected by practitioners due to various factors, such as time and financial constraints. During this step, the questionnaire is evaluated by its potential user population, domain experts, and other practitioners. This allows any problems with the questionnaire to be removed before the critical administration phase. Typically, numerous problems are encountered during the pilot stage, such as errors within the questionnaire, redundant questions, and questions that the participants simply do not understand or find confusing. Wilson and Corlett (1995) recommended that the pilot stage should comprise the following three stages:

- *Individual criticism*: The questionnaire should be administered to several colleagues who are experienced in questionnaire construction, administration, and analysis. These colleagues should be encouraged to offer criticisms of the questionnaire.

- *Depth interviewing*: Once the individual criticisms have been attended to and any changes have been made, the questionnaire should be administered to a small sample of the intended population. Once they have completed the questionnaire, the participants should be subjected to an interview regarding the answers that they provided. This allows the analyst to ensure that the questions were fully understood and that the correct (required) data are obtained.

- *Large-sample administration*: The redesigned questionnaire should then be administered to a large sample of the intended population. This allows the analyst to ensure that the correct data are being collected and also that sufficient time is available to analyze the data. Redundant questions can also be highlighted during this stage. The likely response rate can also be predicted upon the basis of the returned questionnaires in this stage.

Step 5: Questionnaire Administration

Once the questionnaire has been successfully piloted, it is ready to be administered. Exactly how the questionnaire is administered is dependent upon the aims and objectives of the analysis, and also the target population. For example, if the target population can be gathered together at a certain time and place, then the questionnaire could be administered at this time, with the analyst(s) present. This ensures that the questionnaires are completed. However, gathering the target population in one place at the same time can be problematic and so questionnaires have also been administered by post or e-mail. More questionnaires are delivered by way of e-mail, and the provision of an online link directing participants to the questionnaire. There are a number of companies that offer a range of services for the collection and analyses of questionnaire data in this way. Often, however, response rates for questionnaires administered in these ways can be low—often only 10 percent. Procedures to address poor responses rates are available, such as offering payment on completion, the use of reminder e-mails, offering a donation to charity upon return, contacting nonrespondents by e-mail, and sending shortened versions of the initial questionnaire to nonrespondents. All these methods have been shown in the past to improve response rates, but almost all involve extra cost.

Step 6: Data Analysis

Once all (or a sufficient amount) of the questionnaires have been returned or collected, the data-analysis process should begin. This is a lengthy process, the exact time required being dependent upon a number of factors (e.g., number of question items, sample size, required statistical techniques, and data reduction). Questionnaire data are normally computerized and analyzed statistically.

Step 7: Follow-up Phase

Once the data are analyzed sufficiently and conclusions are drawn, the participants who completed the questionnaire should be informed regarding the outcome of the study. This might include a thank-you letter and an associated information pack containing a summary of the research findings.

2.3.5 Advantages

- Questionnaires offer a very flexible way of collecting large volumes of data from large participant samples.
- When the questionnaire is properly designed, the data-analysis phase should be quick and very straightforward.
- Very few resources are required once the questionnaire has been designed.
- Questionnaires are very easy to administer to large numbers of participants.
- Skilled questionnaire designers can use the questions to direct the data collection.

2.3.6 Disadvantages

- Designing, piloting, administering, and analyzing a questionnaire is time-consuming.
- Reliability and validity of questionnaires are questionable.
- The questionnaire design process is taxing, requiring great skill on the analyst's part.
- Typically, response rates are low (around 10 percent for postal and online questionnaires).
- The answers provided in questionnaires are often rushed and noncommittal.
- Questionnaires are prone to a number of different biases, such as prestige bias.
- Questionnaires can offer a limited output.

2.3.7 Flowchart

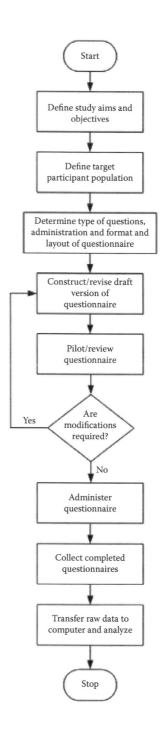

2.3.8 Approximate Training and Application Times

Wilson and Corlett (1995) suggested that questionnaire design is more of an art than a science. Practice makes perfect, and practitioners normally need to make numerous attempts at questionnaire design before becoming proficient at the process (Oppenheim, 2000). Similarly, although the application time associated with questionnaires is at first glance minimal (i.e., the completion phase), when one considers the time expended in the construction and data-analysis phases, it is apparent that the total application time is high.

2.3.9 Reliability and Validity

The reliability and validity of questionnaire techniques are questionable. Questionnaire techniques are prone to a number of biases and often suffer from *social desirability*, whereby the participants are merely "giving the analyst(s) what they want." Questionnaire answers are also often rushed and noncommittal. In a study comparing 12 HF techniques, Stanton and Young (1999) reported that questionnaires demonstrated an acceptable level of inter-rater reliability but unacceptable levels of intrarater reliability and validity.

2.3.10 Tools Needed

Questionnaires can be both paper-based or developed in digital software. Questionnaire design normally requires a computer, along with a word-processing package such as Microsoft Word; there are also a number of survey and questionnaire-development packages available online. In the analysis of questionnaire data, a spreadsheet package such as Microsoft Excel™ is required, and a statistical software package such as SPSS™ is also required to treat the data statistically.

2.3.11 Example

Figure 2.1 shows the start of an online questionnaire developed for gathering data from urban development experts on the topic of infill development of main streets.

A Sociotechnical Systems Approach to the Optimization of Complex Urban Environments

This research focuses on the optimization of (1) Main streets and (2) Multimodal urban corridors.

What is a main street?
Main streets, sometimes known as high streets, are locations typically within a town center, district center, and major or principal center. A main street is considered here to include the areas of the road pavement, road reserve, and the land uses up to 100 m from the road reserve (refer to Figure 1). It is more than just the street itself; it is the area within the street corridor, the interface of pathways, roads, and built form where people and technology converge in a complex urban setting.

What is a multimodal urban corridor?
Multimodal urban corridors are located along higher order transport corridors. They contain a variety of transport options and link two distinct urban areas (town to town or city to suburbs as examples). At a larger scale to a main street, multimodal urban corridors incorporate mass transit functions and a full range of urban land uses. A multimodal urban corridor includes the road reserve and the adjacent land uses up to 800 m away (refer to Figure 1).

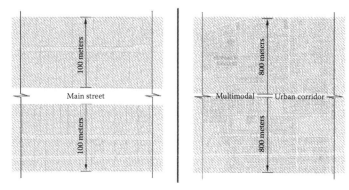

Figure 1: Main street and multimodal urban corridor

You have been identified as a subject matter expert, so please tell us a little more detail about your professional experience.

1. What discipline(s) best describes your professional experience?

☑ Landscape Architecture

☐ Architecture

☐ Transport Planning

Figure 2.1   Introduction to an urban development questionnaire. (From Patorniti, N. P. et al., *Habitat International*, 66, 42–48, 2017.)   (*Continued*)

☑ Urban Design

☐ Strategic Town Planning

☐ Statutory Approvals/Permitting Town Planning

☐ Civil Engineering

☐ Economics

☐ Social Planning

Other (please specify) [   ]

2. How many years have you been practicing in the above discipline(s)? 0

[6–10 years   ▼]

3. What organization do you primarily belong to?

⦿ Consultancy Services Company

○ Nongovernmental Organization

○ Professional Institution

○ Government/The Public Sector

○ Academia

Other (please specify) [   ]

4. Please indicate your level of expertise in main street and multimodal urban corridors (e.g., designing, planning, assessing, teaching, researching).

	Low	Medium	High
Main street	⦿ Main street low	○ Main street medium	○ Main street high
Multimodal urban corridor	⦿ Multimodal urban corridor low	○ Multimodal urban corridor medium	○ Multimodal urban corridor high

Figure 2.1 (Continued)   Introduction to an urban development questionnaire.   (*Continued*)

5. In what country do you currently work?

[▼]

What other countries have you worked in (please specify).

[   ]

Figure 2.1 (Continued)   Introduction to an urban development questionnaire.

2.3.12 Recommended Text(s)

Oppenheim, A. N. (2000). *Questionnaire design, interviewing and attitude measurement*. London: Continuum.

2.4 Observation

2.4.1 Background and Applications

Observational techniques are used to gather data regarding the physical and verbal aspects of a task or scenario. These include tasks catered for by the system, the individuals performing the tasks, the tasks themselves (task steps and sequence), errors made, communications between individuals, the technology used by the system in conducting the tasks (controls, displays, communication technology, etc.), the system environment, and the organizational environment. Observation has been extensively used and typically forms the starting point of an analysis effort. The most obvious and widely used form of observational technique is direct observation, whereby an analyst visually records a particular task or scenario. However, a number of different forms of observation exist, including direct observation as well as participant observation and remote observation. Drury (1990) suggested that there are five different types of information that can be elicited from observational techniques: the sequence of activities, the duration of activities, the frequency of activities, the amount of time spent in states, and spatial movement. Moreover, physical (or visually recorded) data, verbal data are also recorded, in particular verbal interactions between the agents involved in the scenario under analysis. Observational techniques can be used at any stage of the design process to gather information regarding existing or proposed designs.

2.4.2 Domain of Application

Generic.

2.4.3 Application in Land Use Planning and Urban Design

The use of observational study is common in LUP & UD. It is useful for exploring and understanding the array of interactions that different users may have within a particular urban environment.

2.4.4 Procedure and Advice

There is no set procedure for carrying out an observational analysis. The procedure would normally be determined by the nature and scope of analysis required. A typical observational analysis procedure can be split into the following three phases: the observation design stage, the observation application stage, and the data-analysis stage. The following procedure provides the analyst with a general set of guidelines for conducting a *direct*-type observation.

>   *Step 1: Define the Objective of the Analysis*
>   The first step in observational analysis involves clearly defining the aims and objectives of the observation. This should include determining what design or system is under analysis, in which environment the observation will take place, which user groups will be observed, what type of scenarios will be observed, and what data are required. Each point should be clearly defined and stated before the process continues.
>   *Step 2: Define the Scenario(s)*
>   Once the aims and objectives of the analysis are clearly defined, the scenario(s) to be observed should be defined and described further. For example, when conducting an observational analysis of a public space, the type of scenario required should be clearly defined. Normally, the analyst(s)

has a particular type of scenario in mind—for example, the use and users of the space at a particular time of day or night. The exact nature of the required scenario(s) should be clearly defined by the observation team. It is recommended that a hierarchical task analysis is then conducted for the scenario under analysis.

Step 3: Observation Plan

Once the aim of the analysis is defined and the type of scenario to be observed is determined, the analysis team should proceed to plan the observation. The team should consider what they are hoping to observe, what they are observing, and how they are going to observe it. Depending upon the nature of the observation, access to the space, or system, in question should be gained first. This may involve holding meetings with, or gaining approvals from, government organizations or the establishment in question and is typically a lengthy process. Any recording tools should be defined, and the length of observations should also be determined. In addition, placement of video- and audio-recording equipment should be considered. To make things easier, a walkthrough of the system/environment/scenario under analysis is recommended. This allows the analyst(s) to become familiar with the task in terms of the activity conducted, the time taken, location, and also the system under analysis.

Step 4: Pilot Observation

In any observational study, a pilot or practice observation is crucial. This allows the analysis team to assess the quality and any problems with the data collection, such as noise interference or problems with the recording equipment. If major problems are encountered, the observation may have to be redesigned. Steps 1–4 should be repeated until the analysis team is happy that the quality of the data collected will be sufficient for their study requirements.

Step 5: Conduct Observation

Once the observation has been designed, the team should proceed with the observation(s). Typically, data are recorded using video- and audio-recording equipment. An observation transcript is also created during the observation. An example of an observation transcript template is presented in Table 2.2. Observation length and timing are dependent upon the scope and requirements of the analysis and also the scenario(s) under analysis. The observation should end only when the required data are collected.

Step 6: Data Analysis

Once the observation is complete, the data-analysis procedure begins. Typically, the starting point of the analysis phase involves typing up the observation notes or transcript made during the observation. This is a very time-consuming process but is crucial to the analysis. Depending upon the analysis requirements, the team should then proceed to analyze the data in the format that is required, such as frequency of actions, verbal interactions, and sequence of use. When analyzing visual data, typical user behaviors are coded into specific groups. The software package *Observer XT* is frequently used to aid the analyst in this process.

Step 7: Further Analysis

Once the initial process of transcribing and coding the observational data is complete, further analysis of the data begins. Depending upon the nature of the analysis, observation data are used to inform a number of different analyses, such as task analysis, error analysis, and communications analysis.

Step 8: Participant Feedback

If applicable, once the data have been analyzed and conclusions have been drawn, the participants involved should be provided with feedback of some sort. This could be in the form of a feedback session or a letter to each participant. The type of feedback used is determined by the analysis team (Table 2.2).

Table 2.2   Example of an Observation Transcript Template

TIME	ACTOR	EVENT/CONVERSATION	DECISION	CONTEXT	TAG NAME
08.00	PS	Entered the park		First interface	PS01
08.05	PS	Sat on exposed bench facing east	Sit in the morning sun	Settlement	PS01
08.06	OD	Entered park, traversed on pathway	Transit	Commuting through	OD01
08.15	BD	Entered park with dog			BD01
08.16	BD	Exploring open grassed area with dog	Use for animal amenity	Exercise/amenity	BD02

2.4.5 Flowchart

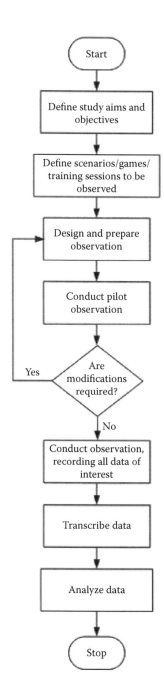

2.4.6 Advantages

- Observational data provide a *real-life* insight into the activity performed in complex systems.
- Various data can be elicited from an observational study, including number of users of a space—such as pedestrian or traffic counts and types, as a well as the functionality and usability of a space.
- Observation has been used extensively in a wide range of domains.
- Observation provides objective information.
- Detailed physical-task performance data are recorded, including social interactions and any environmental influences (Kirwan and Ainsworth, 1992).
- Observation analysis can be used to highlight problems with existing operational systems. It can be used in this way to inform the design of new systems or designs.
- Specific scenarios are observed in their *real-world* setting.
- Observation is typically the starting point in any HF analysis effort, and observational data are used as the input into numerous HF analyses techniques, such as error identification techniques, task analysis, communications analysis, and charting techniques.

2.4.7 Disadvantages

- Observational techniques are intrusive to task performance.
- Observation data are prone to various biases. Knowing that they are being watched tends to elicit new and different behaviors in participants. For example, people may not use the space in *illegal* ways—drinking, smoking, skating, and so on.
- Observational techniques are time-consuming in their application, particularly the data-analysis procedure. Kirwan and Ainsworth (1992) suggest that when conducting the

transcription process, 1 hour of recorded audio data takes an analyst approximately 8 hour to transcribe.

- Cognitive aspects of the task under analysis are not elicited using observational techniques. Verbal protocol analysis is more suited for collecting data on the cognitive aspects of task performance.
- An observational study can be both difficult and expensive to set up and conduct. Gaining access and permissions is often extremely difficult and very time-consuming. Observational techniques are also costly, as they require the use of expensive recording equipment (digital video camera, audio-recording devices).
- Causality is a problem. Errors can be observed and recorded during an observation, but why the errors occur may not always be clear.
- The analyst has only a limited level of experimental control.
- In most cases, a team of analysts is required to perform an observation study. It is often difficult to acquire a suitable team with sufficient experience in conducting observational studies.

2.4.8 Related Methods

There are a number of different observational techniques, including indirect observation, participant observation, and remote observation. The data derived from observational techniques are used as the input to a plethora of HF techniques, including task analysis, cognitive task analysis, charting, and human error-identification techniques.

2.4.9 Approximate Training and Application Times

Although the training time for an observational analysis is low (Stanton and Young, 1999), the application time is typically high. The data-analysis phase in particular is extremely time-consuming and so too is the transcription phase, as highlighted earlier by Kirwan and Ainsworth (1992).

2.4.10 Reliability and Validity

Observational analysis is beset by a number of problems that can potentially affect the reliability and validity of the technique. According to Baber and Stanton (1996), problems with causality, bias (in a number of forms), construct validity, external validity, and internal validity can all arise unless the correct precautions are taken. Although observational techniques possess a high level of face validity (Drury, 1990) and ecological validity (Baber and Stanton, 1996), analyst or participant bias can adversely affect their reliability and validity.

2.4.11 Tools Needed

For a thorough observational analysis, the appropriate visual- and audio-recording equipment is necessary. Simplistic observational studies can be conducted using pen and paper only. However, for observations in complex, dynamic systems, more sophisticated equipment is required, such as video- and audio-recording equipment. For the purposes of data analysis, a computer with the *Observer XT* software is required.

2.4.12 Recommended Text(s)

Baber, C., & Stanton, N. A. (1996b). Observation as a technique for usability evaluation. In P. W. Jordan, B. Thomas, B. A. Weerdmeester, & I. McClelland (Eds.), *Usability evaluation in industry* (pp. 85–94). London: Taylor & Francis Group.

Drury, C. (1990). Methods for direct observation of performance. In J. R. Wilson & E. N. Corlett (Eds.), *Evaluation of human work*, (pp. 45–68). London: Taylor & Francis Group.

3

TASK ANALYSIS METHODS

3.1 Introduction

Although data collection techniques are used to collect specific data regarding the activity performed in complex systems, task-analysis techniques describe and represent the activity performed. Task-analysis techniques, can be used to understand and represent human and system performance in a particular land use planning and urban design (LUP & UD) task or scenario under analysis. Task analysis involves identifying tasks, collecting task data, analyzing the data so that tasks are understood, and then producing a documented representation of the analyzed tasks (Annett et al., 1971). According to Diaper and Stanton (2004), there are more than 100 task-analysis techniques described in the literature.

Typical task-analysis techniques are used to break down tasks or scenarios into component task steps or physical operations. According to Kirwan and Ainsworth (1992), task analysis can be defined as the study of what an operator (person) is required to do (their actions and cognitive processes) to achieve system goals. In an LUP & UD sense, this could be associated with systems of work, such as development assessment or site analysis processes, or indeed how an individual interacts with a particular urban environment.

The use of task-analysis techniques is widespread, with applications in a wide range of domains, including military operations, aviation (Stanton et al., 2016), air-traffic control, driving (Walker et al., 2001), public technology (Stanton and Stevenage, 1998), product design, and nuclear petrochemical domains, to name a few. Diaper and Stanton (2004) suggested that task analysis is potentially the most powerful technique available to human–computer interaction (HCI) practitioners. Although Stanton (2006) also suggested that task analysis is the central method for the design and analysis of system performance, involved in everything from design concept to system development and operation. Stanton also highlighted the role of task analysis in task allocation, procedure design, training design, and interface design.

A task analysis of the task(s) and system under analysis is the next logical step after the data collection process. Specific data are used to conduct a task analysis, allowing the task to be described in terms of the individual task steps required, the technology used in completing the task (controls, displays, etc.), and the sequence of the task steps involved. The task description offered by task-analysis techniques is then typically used as the input to further analysis techniques, such as error identification (EI) techniques and process charting techniques. For example, systematic human-error reduction and prediction approach (SHERPA) (Embrey, 1986) and the human-error template (Marshall et al., 2003) are both EI techniques (Chapter 5) that are applied to the bottom-level task steps identified in a hierarchical task analysis (HTA). In doing so, the task under analysis can be scrutinized to identify potential errors that might occur during the performance of that task.

The popularity of task-analysis techniques is a direct function of their usefulness and flexibility. Typically, a task analysis of some sort is required in any HF analysis effort, be it usability evaluation, error identification, or performance evaluation. Task-analysis outputs are particularly useful, providing a step-by-step description of the activity under analysis. Moreover, analysts using task-analysis approaches often develop a (required) deep understanding of the activity under analysis.

However, task-analysis techniques are not without their flaws. The resource usage incurred when using such approaches is often considerable. The data collection phase is time-consuming and often requires the provision of video- and audio-recording equipment. Such techniques are also typically time-consuming in their application, and many reiterations are needed before an accurate representation of the activity under analysis is produced. Task-analysis techniques also suffer from reliability problems, and different analysts may produce entirely different representations of the same activity. Similarly, analysts may produce different representations of the same activity on different occasions.

There are a number of different approaches to task analysis available to the LUP & UD practitioner; however, the most commonly used and well-known task-analysis technique is HTA (Annett, 2004; Stanton, 2006). HTA involves breaking down the task under analysis into a nested hierarchy of goals, operations, and plans. Verbal protocol analysis (VPA) is used to derive the processes (cognitive and physical) that an individual uses to perform a task. VPA involves creating a written transcript of an individuals behavior as they perform the task under analysis. Task decomposition (Kirwan and Ainsworth, 1992) can be used to create a detailed task description using specific categories to exhaustively describe actions, goals, controls, error potential, and time constraints.

Task-analysis techniques have evolved in response to increased levels of complexity and the increased use of teams within work settings. A wide variety of task-analysis procedures now exist, including techniques designed to consider the cognitive aspects of decision-making and activity in complex systems (cognitive task analysis— Chapter 4). Cognitive task-analysis techniques, such as the critical decision method (Klein and Armstrong, 2004), use probe interview techniques to analyze, understand, and represent the unobservable cognitive processes associated with tasks or work. In this chapter, five task-analysis techniques will be explored: (1) HTA, (2) VPA, (3) Task decomposition, (4) Operator sequence diagrams (OSD), and (5) Critical path analysis (CPA).

3.2 Hierarchical Task Analysis

3.2.1 Background and Applications

HTA (Annett, 2004; Stanton, 2006) is the most popular task-analysis technique and has become perhaps the most widely used of all available HF techniques. Originally developed in response to the need for greater understanding of cognitive tasks (Annett, 2004), HTA involves describing the activity under analysis in terms of a hierarchy of goals, sub-goals, operations, and plans. The end result is an exhaustive description of task activity. One of the main reasons for the enduring popularity of the technique is its flexibility and the scope for further analysis that it offers to the practitioner.

The majority of HF analysis methods either require an initial HTA of the task under analysis as their input or at least are made significantly easier through the provision of an HTA. HTA acts as an input into numerous HF analyses techniques, such as EI, allocation of function, workload assessment, interface design and evaluation, and many more. Consequently, HTA has been applied across a wide spectrum of domains, including the process control and power generation industries (Annett, 2004), emergency services, military applications (Kirwan and Ainsworth, 1992), civil aviation (Marshall et al., 2003), driving (Walker et al., 2001), public technology (Stanton and Stevenage, 1998), and retail (Shepherd, 2002), to name a few.

3.2.2 Domain of Application

HTA was originally developed for the chemical processing and power generation industries (Annett, 2004). However, the technique is generic and can be applied in any domain.

3.2.3 Application in Land Use Planning and Urban Design

Although there are no published applications of the HTA method in an LUP & UD context, it is clearly suited to the analyses of LUP & UD tasks. HTA deals with the goal, objective, action-related physical, and cognitive task performance that is already a familiar approach in LUP & UD.

3.2.4 Procedure and Advice

Step 1: Define Task(s) Under Analysis
The first step in conducting an HTA is to clearly define the task (or tasks) under analysis. Moreover, identifying the task under analysis, the purpose of the task-analysis effort should also be defined.

Step 2: Data Collection Process
Once the task (or tasks) under analysis is clearly defined, specific data regarding the task should be collected. The data collected during this process are used to inform the development

of the HTA. Data regarding the task steps involved, the technology used, interaction between humans and machines or objects in the environment, team members, decision-making, and task constraints should be collected. There are a number of ways to collect these data, including observations, interviews with subject matter experts, questionnaires, and walkthroughs. The techniques used are dependent upon the analysis effort and the various constraints imposed, such as time and access constraints. Once sufficient data regarding the task under analysis are collected, the development of the HTA should begin.

Step 3: Determine the Overall Goal of the Task

The overall goal of the task under analysis should first be specified at the top of the hierarchy, that is, "Undertake a site analysis," "Assess a development application" even as simple as "Boil kettle" or "Listen to in-car entertainment" (Stanton and Young, 1999).

Step 4: Determine Task Sub-goals

Once the overall task goal has been specified, the next step is to break this overall goal down into meaningful sub-goals (usually four or five, but this is not rigid), which together form the tasks required to achieve the overall goal. In the task *Assess a development application*, this is broken down into the following sub-goals: *Application received by development application authority*, *Assessment manager reviews application*, Application is referred to, *Information is requested*, *Public is notified*, and *Decide on application*. In an HTA of a Ford in-car radio (Stanton and Young, 1999), the overall task goal *Listen to in-car entertainment* was broken down into the following sub-goals *Check unit status*, *Press on/off button*, *Listen to the radio*, *Listen to cassette*, and *Adjust audio preferences*.

Step 5: Sub-goal Decomposition

Next, the analyst should break down the sub-goals identified during step four into further sub-goals and operations, according to the task step in question. This process should go on until an appropriate operation is reached. The bottom level of any branch in an HTA should always be an operation. Although everything above an operation specifies goals, operations actually say what needs to be done. Therefore, operations are actions to be made by an agent to achieve the associated goal. For example, in the HTA of the planning task *Assess a development application*, the sub-goal *Application is referred to* is broken down into the following operations: *Referral agency reviews application* and *Referral agency gives confirmation*.

Step 6: Analysis of Plans

Once all the sub-goals and operations have been fully described, the plans need to be added. Plans dictate how the goals are achieved. A simple plan would say do 1, then 2, and then 3. Once the plan is completed, the agent returns to the superordinate level. Plans do not have to be linear and exist in many forms, such as do 1, or 2 and 3. The different types of plans used are presented in Table 3.1. The output of an HTA can either be a tree diagram (Figure 3.1) or a tabular diagram (Table 3.2).

3.2.5 Advantages

- HTA requires minimal training and is easy to implement.
- The output of an HTA is extremely useful and forms the input for numerous HF analyses, such as error analysis, interface design, and evaluation and allocation of function analysis.
- It is an extremely flexible technique that can be applied in any domain for a variety of purposes.
- It is quick to use in most instances.

Table 3.1   Example HTA Plans

PLAN	EXAMPLE
Linear	Do 1, then 2, then 3
Nonlinear	Do 1, 2, and 3 in any order
Simultaneous	Do 1, then 2 and 3 at the same time
Branching	Do 1; if X present, then do 2 then 3; but if X is not present, then exit
Cyclical	Do 1, then 2, then 3 and repeat until X
Selection	Do 1, then 2 or 3

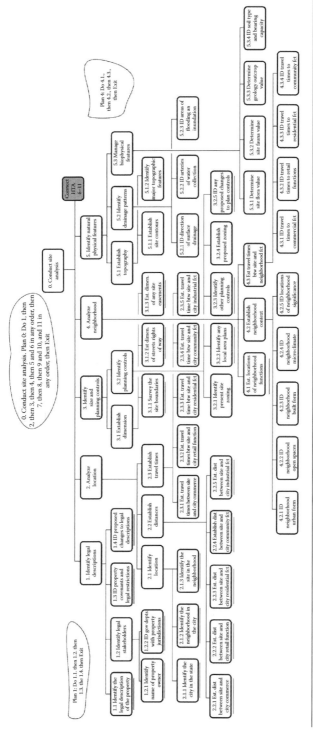

Figure 3.1 Part of Site analysis HTA tree diagram. (Adapted from White, E., *Site Analysis: Diagramming Information for Architectural Design*, Architectural Media, Florida, 1983.) *(Continued)*

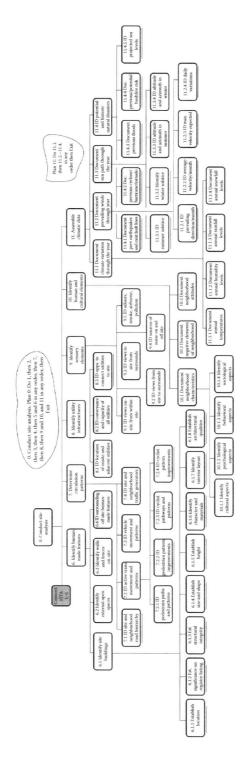

Figure 3.1 (Continued) Site analysis HTA tree diagram.

Table 3.2   Extract from Site Analysis HTA Plan (Full HTA Plan at the End of This Chapter, Table 3.6 page 96)

---

0. **Conduct site analysis**
   Plan 0: Do 1, then 2, 3, 4, 5, and 6 in any order, then 7, 8, 9, 10, 11, in any order, then EXIT

1. Identify legal descriptions (Plan 1: Do 1.1, then 1.2, 1.3, 1.4, respectively, then EXIT)
   1.1. Identify the legal description of the property
   1.2. Identify legal stakeholders (Plan 1.2: Do 1.2.1, then 1.2.2, then EXIT)
       1.2.1. Identify name of property owner
       1.2.2. Identify government departments with property jurisdictions
   1.3. Identify property covenants and legal restrictions
   1.4. Identify any projected or proposed changes to legal descriptions
2. Analyze Location (Plan 2: Do 2.1, followed by 2.2 and 2.3, then EXIT)
   2.1. Identify location (Plan 2.1: Do 2.1.1, followed by 2.1.2 and 2.1.3, then EXIT)
       2.1.1. Identify the city in the state
       2.1.2. Identify the neighborhood in the city
       2.1.3. Identify the site in the neighborhood
   2.2. Establish distances (Plan 2.2: Do 2.2.1–2.2.5 in any order, then EXIT)
       2.2.1. Establish the distance between the site and the city commercial functions
       2.2.2. Establish the distance between the site and the city retail functions
       2.2.3. Establish the distance between the site and the city residential functions
       2.2.4. Establish the distance between the site and the city community functions
       2.2.5. Establish the distance between the site and the city industrial functions
   2.3. Establish travel times (Plan 2.3: Do 2.3.1–2.3.5 in any order, then EXIT)
       2.3.1. Est. travel times between the site and the city commercial functions (Plan 2.3.1: Do 2.3.1.1–2.3.1.4 in any order, then EXIT)
           2.3.1.1. Est. walking travel time
           2.3.1.2. Est. cycling travel time
           2.3.1.3. Est. public transport travel time
           2.3.1.4. Est. private vehicle travel time
       2.3.2. Est. travel times between the site and the city retail functions (Plan 2.3.2: Do 2.3.2.1–2.3.2.4 in any order, then EXIT)
           2.3.2.1. Est. walking travel time
           2.3.2.2. Est. cycling travel time
           2.3.2.3. Est. public transport travel time
           2.3.2.4. Est. private vehicle travel time
       2.3.3. Est. travel times between the site and the city residential functions (Plan 2.3.3: Do 2.3.3–2.3.3.4 in any order, then EXIT)
           2.3.3.1. Est. walking travel time
           2.3.3.2. Est. cycling travel time
           2.3.3.3. Est. public transport travel time
           2.3.3.4. Est. private vehicle travel time
       2.3.4. Est. travel times between the site and the city community functions (Plan 2.3.4: Do 2.3.4.1–2.3.4.4 in any order, then EXIT)
           2.3.4.1. Est. walking travel time
           2.3.4.2. Est. cycling travel time

(*Continued*)

Table 3.2 (*Continued*)   Extract from Site Analysis HTA Plan (Full HTA Plan at the End of This Chapter, page x)

2.3.4.3. Est. public transport travel time
2.3.4.4. Est. private vehicle travel time
2.3.5. Est. travel times between the site and the city industrial functions (Plan 2.3.5: Do 2.3.5.1–2.3.5.4 in any order, then EXIT)
2.3.5.1. Est. walking travel time
2.3.5.2. Est. cycling travel time
2.3.5.3. Est. public transport travel time
2.3.5.4. Est. private vehicle travel time

Source: Adapted from White, E., *Site Analysis: Diagramming Information for Architectural Design*, Architectural Media, Florida, 1983.

- The output provides a comprehensive description of the task under analysis.
- It has been used extensively in a wide range of contexts.
- Conducting an HTA gives the user considerable insight into the task under analysis. Salmon et al. (2010) proposed that the process of creating an HTA enables key insights to be gained in addition to the results of the analysis.
- It is an excellent technique to use when requiring a task description for further analysis. If performed correctly, the HTA should depict everything that needs to be done to complete the task in question.
- The technique is generic and can be applied to any task in any domain.
- Tasks can be analyzed to any required level of detail, depending on the purpose.

3.2.6 Disadvantages

- HTA provides mainly descriptive information rather than analytical information.
- It contains little that can be used directly to provide design solutions.
- It does not cater for the cognitive components of the task under analysis.
- The technique may become laborious and time-consuming to conduct for large, complex tasks. The initial data collection

phase is time-consuming and requires the analyst to be competent in a variety of HF techniques, such as interviews, observations, and questionnaires.

- The reliability of the technique may be questionable in some instances. For example, for the same task, different analysts may produce very different task descriptions.
- Conducting an HTA is more of an art than a science, and much practice is required before an analyst becomes proficient in the application of the technique.
- An adequate software version of the technique is yet to emerge.
- There are few prescriptive guidelines on how to apply it (Stanton, 2006).

3.2.7 Related Methods

HTA is widely used in HF and often forms the first step in a number of analyses, such as EI, human reliability assessment (HRA), and mental workload assessment. Stanton (2006) conducted a comprehensive review outlining a variety of applications of HTA, including interface design, error prediction, workload analysis, team performance assessment, and training requirement identification.

Mills (2007) argued that HTA is at its best when used alongside other methods; for example, Mills (2007) used HTA alongside usability context analysis, and Salmon et al. (2010) have developed a piece of software that integrates HTA with additional HF methods, including methods providing insights into workload (the NASA Task Load Index) and error (SHERPA).

3.2.8 Approximate Training and Application Times

According to Annett (2004), a study by Patrick et al. (2000) gave students a few hours training with not entirely satisfactory results on the analysis of a very simple task, although performance improved with further training. A survey by Ainsworth and Marshall (1998) found that the more experienced practitioners produced more complete and acceptable analyses.

Stanton and Young (1999) report that the training and application time for HTA is substantial. The application time associated with HTA is dependent upon the size and complexity of the task under analysis. For large, complex tasks, the application time for HTA would be high. Salmon et al. (2010) also suggested that HTA application times are high, stating that the high fidelity of information captured in an HTA can increase application times to almost double those of other methods such as cognitive work analysis (Chapter 10). However, they also argue that increased application times are correlated with a greater granularity of detail, suggesting that if cognitive work analysis was conducted to the same level of granularity, its application time would be far higher than HTA.

3.2.9 Reliability and Validity

According to Annett (2004), the reliability and validity of HTA are not easily assessed. From a comparison of 12 HF techniques, Stanton and Young (1999) reported that the technique achieved an acceptable level of validity but a poor level of reliability. The reliability of the technique is certainly questionable. It seems that different analysts with different levels of experience may produce entirely different analyses for the same task (intra-analyst reliability). Similarly, the same analyst may produce different analyses on different occasions for the same task (interanalyst reliability).

3.2.10 Tools Needed

HTA can be carried out using pencil and paper only. The HTA output can be developed and presented in a number of software applications, such as Microsoft Visio, Microsoft Word, and Microsoft Excel. A number of HTA software tools also exist, such as the C@STTA HTA tool.

3.2.11 Example

Figure 3.1 and 3.2 provide an example of HTA.

3.2.12 Flowchart

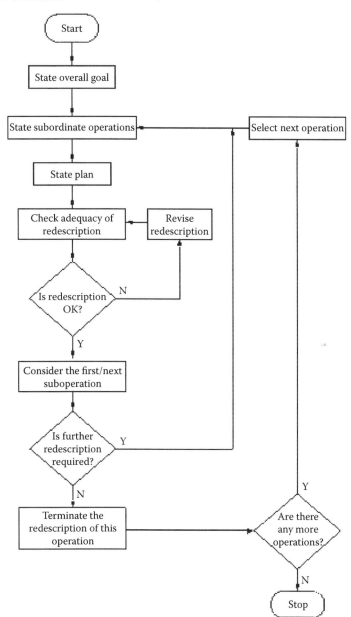

3.3 Verbal Protocol Analysis

3.3.1 Background and Applications

VPA is used to derive descriptions of the processes (cognitive and physical) that an individual uses to perform a task. It involves creating a written transcript of behavior as they perform the task or scenario under analysis. The transcript is based upon the individual *thinking aloud* as he or she conducts the task under analysis. VPA has been used extensively as a means of gaining an insight into the cognitive aspects of complex behaviors. It has been used in many domains, ranging from investigating expertise in nursing (Hoffman et al., 2009), to exploring the reliability of personality questionnaires (Robie et al., 2007), and the examination of the compatibility of motorcyclists' and car drivers' mental representations of the road (Salmon et al., 2014a).

3.3.2 Domain of Application

Generic.

3.3.3 Application in Land Use Planning and Urban Design

VPA can provide detailed insights into a range of important tasks within our built environments. It can capture the experiences of the users of differing abilities within urban systems, such as the walkability of a particular setting; and indeed the quality of wayfinding and legibility cues within complex urban systems.

3.3.4 Procedure and Advice

The following procedure is adapted from Walker (2004).

Step 1: Define Scenario Under Analysis
First, the scenario under analysis should be clearly defined. It is recommended that an HTA is used to describe the task under analysis.

Step 2: Instruct/Train the Participant
Once the scenario is clearly defined, the participant should be briefed regarding what is required of them during the analysis.

What they should report verbally is clarified here. According to Walker (2004), it is particularly important that the participant is informed that they should continue talking even when what they are saying does not appear to make much sense. A small demonstration should also be given to the participant at this stage. A practice run may also be undertaken, although this is not always necessary.

Step 3: Begin Scenario and Record Data

The participant should begin to perform the scenario under analysis. The whole scenario should be audio recorded (at least) by the analyst. It is also recommended that a video recording be made.

Step 4: Verbalization of Transcript

Once collected, the data should be transcribed into a written form. An Excel spreadsheet is normally used. This aspect of VPA is particularly time-consuming and laborious.

Step 5: Encode Verbalizations

The verbal transcript (written form) should then be categorized or coded. Depending upon the requirements of the analysis, the data are coded into one of the following five categories: words, word senses, phrases, sentences, or themes. The encoding scheme chosen should then be encoded according to a rationale determined by the aims of the analysis. The analyst should also develop a set of written instructions for the encoding scheme. These instructions should be strictly adhered to and constantly referred to during the encoding process (Walker, 2004). Once the encoding type, framework, and instructions are completed, the analyst should proceed to encode the data. Various computer software packages are available to aid the analyst with this process, such as General Enquirer and NVivo.

Step 6: Devise Other Data Columns

Once the encoding is complete, the analyst should devise any *other* data columns. This allows the analyst to note any mitigating circumstances that may have affected the verbal transcript.

Step 7: Establish Inter- and Intrarater Reliability

Reliability of the encoding scheme then has to be established (Walker, 2004). In VPA, reliability is established through reproducibility, that is, independent raters need to encode previous analyses.

Step 8: Perform Pilot Study

The protocol analysis procedure should now be tested within the context of a small pilot study. This will demonstrate whether the verbal data collected are useful, whether the encoding system works, and whether inter- and intrarater reliability are satisfactory. Any problems highlighted through the pilot study should be refined before the analyst conducts the VPA for real.

Step 9: Analyze Structure of Encoding

Finally, the analyst can study the results from the VPA. During any VPA analysis, the responses given in each encoding category require summing, and this is achieved simply by adding up the frequency of occurrence noted in each category. Walker (2004) suggests that for a more fine-grained analysis, the structure of encodings can be analyzed contingent upon events that have been noted in the *other data* column(s) of the worksheet or in light of other data that have been collected simultaneously.

3.3.5 Example

The following example, in Figure 3.2, represents a transcription and encoding sheet from a driving study (Walker, 2004) that considered the impacts of adjacent land use planning on the driving task. Similar sheets can be used for the detailed organization and analysis of any VPA data.

In addition, the transcribed data may be analyzed utilizing a range of coding methods or software such as *NVivo*. The example in Figure 3.3 shows the analysis of transcribed verbal protocol data from a walkability task in a busy urban environment (Pratt, 2017). Here the software *Leximancer* has been used to identify key concepts from the data set.

3.3.6 Advantages

- VPA provides a rich data source.
- It is particularly effective when used to analyze sequences of activities.

TIME	VERBALIZATIONS	ENCODING											EVENTS
		BEHAV.				COG.			F/B				
mm:ss		OB	BC	RE	OT	PC	CM	PR	AC	SD	CD	IN	
01:34	70 mph, 5th gear		1			1					1	1	Glances at gear lever
01:36	2800 rpm		1			1					1		
01:38	that is quite smooth		1				1						
01:40	he's slowing down			1		1							Other car crossing from lane 3 over
01:42	don't know what is wrong with him	1											to hand shoulder in front of driver
01:44													
01:46													
01:48													
01:50													
01:52													
01:54													
01:56													
01:58													
02:00													
02:02													
02:04	It's all clear ahead			1		1							
02:06													
02:08	chap behind has eased off a bit luckily			1		1	1						
02:10													
02:12	make my intention clear that I'm going right	1							1				Indicating right
02:14	so I'll stick to the right side of this lip lane	1	1				1						
02:16													
02:18													
02:20	bit worried about overtaking him	1			1		1						passing other vehicle
02:22													
02:24													
02:26													
	Section frequency counts	4	3	1	5	5	5	0	1	0	1	2	
02:28													

Figure 3.2   Example transcription and encoding sheet from a walkability task.

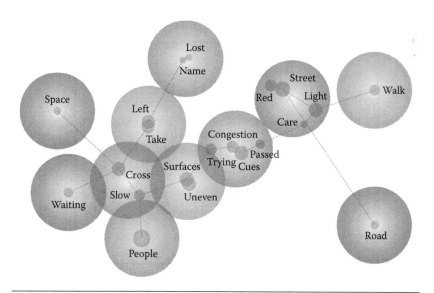

Figure 3.3   Leximancer concept identification of verbal protocol data.

- Verbalizations can provide a genuine insight into cognitive processes.
- Domain experts can provide excellent verbal data.
- It has been used extensively in a wide variety of domains.
- It is simple to conduct with the right equipment.

3.3.7 Disadvantages

- Data analysis (encoding) can become extremely laborious and time-consuming.
- VPA is a very time-consuming method to apply (data collection and data analysis).
- It is difficult to verbalize cognitive behavior. Researchers have been cautioned in the past for relying on verbal protocol data (Militello and Hutton, 2000).
- Verbal commentary can sometimes serve to change the nature of the task.
- Complex tasks involving high demand can often lead to a reduced quantity of verbalizations (Walker, 2004).
- Strict procedure is often not adhered to fully.
- VPA is prone to bias on the participant's behalf.

3.3.8 Related Methods

VPA is related to observational techniques such as walkthroughs and direct observation. Task-analysis techniques such as HTA are often used in constructing the scenario under analysis. VPA is frequently used alongside other HF methods; for example, Hoffman et al. (2009) employed both VPA and retrospective interviewing in their exploration of the differences in decision-making between experts and novices within the medical nursing domain. The method can also act as an input for numerous other HF methods; for example, Walker et al. (2014) used verbal protocols as an input to an automatic concept map tool to develop illustrations of car drivers' and motorcyclists' mental representations of driving.

3.3.9 *Approximate Training and Application Times*

Although the technique is very easy to train, the VPA procedure is time-consuming to implement. According to Walker (2004), if transcribed and encoded by hand, 20 min of verbal transcript data at around 130 words per minute can take between 6 and 8 hours to transcribe and encode.

3.3.10 *Reliability and Validity*

Walker (2004) reported that the reliability of the technique is reassuringly good. For example, Walker et al. (2001) used two independent raters and established inter-rater reliability at Rho = 0.9 for rater 1 and Rho = 0.7 for rater 2. Intrarater reliability during the same study was also high, being in the region of Rho = 0.95.

Hoffman et al. (2009) argued that the use of both concurrent and retrospective protocols increases the reliability and validity of the method as one measures long-term memory and the other short-term memory. They proposed that using researchers with domain experience and maintaining consistency of researchers throughout analysis also increase levels of reliability and consistency.

3.3.11 *Tools Needed*

A VPA can be conducted using pen and paper, a digital audio-recording device, and a video recorder if required. The device or system under analysis is also required. For the data-analysis part of VPA, Excel is normally required, although this can be done using pen and paper. A number of software packages can also be used by the analyst, including *Observer XT*, General Enquirer, TextQuest, NVivo, and Leximancer.

3.3.12 Flowchart

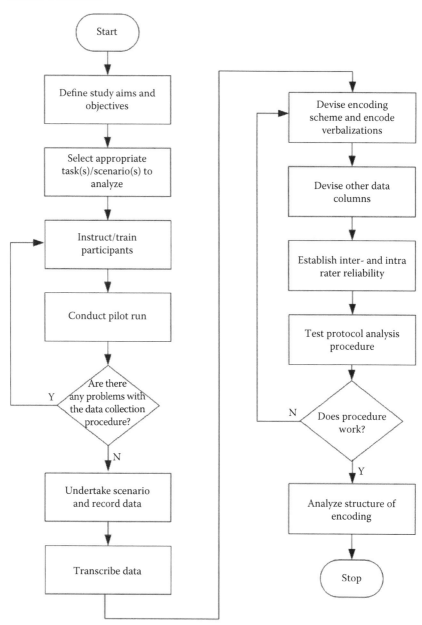

3.3.13 Recommended Text(s)

Bainbridge, L. (1995). Verbal protocol analysis. In J. R. Wilson & E. N. Corlett (Eds.) *Evaluation of human work: A practical ergonomics methodology* (pp. 161–179). London: Taylor & Francis Group.
Noyes, J. M. (2006). Verbal protocol analysis. In W. Karwowski (Ed.), *International encyclopaedia of ergonomics and human factors* (2nd ed.), (pp. 3390–3392). London: Taylor & Francis Group.
Walker, G. H. (2004). Verbal protocol analysis. In N. A. Stanton et al., (Eds.), *The handbook of human factors and ergonomics methods* (pp. 30.1–30.8). Boca Raton, FL: CRC Press.

3.4 Task Decomposition

3.4.1 Background and Applications

Kirwan and Ainsworth (1992) described the task-decomposition methodology that can be used to gather detailed information regarding a particular task or scenario. Task decomposition involves describing the task or activity under analysis and then using specific task-related information to decompose it in terms of statements regarding the task. The task can be broken down to describe a variety of task-related features, including the objects and interface components used (e.g., signage, seating), the time taken, errors made, feedback, and decisions required to undertake the task. The categories used to decompose the task steps should be chosen by the analyst based on the requirements of the analysis. There are numerous decomposition categories that can be used and new categories can be developed if required by the analysis. According to Kirwan and Ainsworth (1992), Miller (1953) was the first practitioner to use the task-decomposition technique.

Miller recommended that each task step should be decomposed around the following categories:

- Description
- Subtask
- Cues initiating action
- Controls used
- Decisions
- Typical errors
- Response

- Criterion of acceptable performance
- Feedback

However, further decomposition categories have since been defined (e.g., Kirwan and Ainsworth, 1992). It is recommended that the analyst develops a set of decomposition categories based upon the analysis requirements.

3.4.2 Domain of Application

Generic.

3.4.3 Application in Land Use Planning and Urban Design

The range and variety of complex tasks undertaken by both the users of urban and regional environments, and the complex procedures utilized to assess and design cities and towns render this method infinitely useful for LUP & UD.

3.4.4 Procedure and Advice

Step 1: HTA
   The first step in a task-decomposition analysis involves creating an initial description of the task or scenario under analysis. It is recommended that an HTA is conducted for this purpose, as a goal-driven, step-by-step description of the task is particularly useful when conducting a task-decomposition analysis.

Step 2: Create Task Descriptions
   Once an initial HTA for the task under analysis has been conducted, the analyst should create a set of clear task descriptions for each of the different task steps. These descriptions can be derived from the HTA developed during Step 1. The task description should give the analyst enough information to determine exactly what has to be done to complete each task element. The detail of the task descriptions should be determined by the requirements of the analysis.

Step 3: Choose Decomposition Categories
   Once a sufficient description of each task step is created, the analyst should choose the appropriate decomposition categories. Kirwan

and Ainsworth (1992) suggested that there are three types of decomposition categories: descriptive, organization-specific, and modeling. Table 3.3 presents a taxonomy of descriptive decomposition categories that have been used in various studies.

Step 4: Information Collection

Once the decomposition categories have been chosen, the analyst should create a data collection pro forma for each decomposition category. The analyst should then work through each decomposition category, recording task descriptions and gathering the additional information required for each of the decomposition headings. To gather this information, Kirwan and Ainsworth (1992) suggested that there are many possible methods to use, including observation, system documentation, procedures, training manuals, and discussions with system personnel and designers. Interviews, questionnaires, VPA, and walkthrough analysis can also be used.

Step 5: Construct Task Decomposition

The analyst should then put the data collected into a task-decomposition output table. The table should comprise all

Table 3.3 Task-Decomposition Categories

TASK-DECOMPOSITION CATEGORIES		
TASK DESCRIPTION	CRITICAL VALUES	SUBTASKS
Activity/behavior type	Job aids required	Communications
Task/action verb	Actions required	Coordination requirements
Function/purpose	Decisions required	Concurrent tasks
Sequence of activity	Responses required	Task outputs
Requirements for undertaking task	Task complexity	Feedback
Initiating cue/event	Task criticality	Consequences
Information	Amount of attention required	Problems
Skills/training required	Performance on task	Likely/typical errors
Personnel requirements/manning	Time taken/time permitted	Errors made
Hardware features	Required speed	Error consequences
Location	Required accuracy	Adverse conditions
Controls used	Criterion of response adequacy	Hazards
Displays used	Other activities	

Source: Kirwan, B. and Ainsworth, L. K., *A Guide to Task Analysis*, Taylor & Francis Group, London, UK, 1992.

of the decomposition categories chosen for the analysis. The amount of detail included in the table is also determined by the scope of the analysis.

3.4.5 Advantages

- Task decomposition is a very flexible approach. By selecting which decomposition categories to use, the analyst can determine the direction and focus of the analysis.
- A task-decomposition analysis has the potential to provide a very comprehensive analysis of a particular task.
- Task-decomposition techniques are easy to learn and use.
- The method is generic and can be used in any domain.
- It provides a much more detailed description of tasks than traditional task-analysis techniques do.
- As the analyst has control over the decomposition categories used, potentially any aspect of a task can be evaluated. In particular, the technique could be adapted to assess the cognitive components associated with tasks (goals, decisions, and situation awareness).

3.4.6 Disadvantages

- As the task technique is potentially so exhaustive, it is a very time-consuming technique to apply and analyze. The HTA only serves to add to the lengthy application time. Furthermore, obtaining information about the tasks (observation, interview, etc.) creates even more work for the analyst.
- Task decomposition can be laborious to perform, involving observations, interviews, and so on.

3.4.7 Example

A task decomposition was performed on the sub-goal "Identify vehicle movement and patterns," Table 3.4, taken from the larger "Site Analyses" HTA (Page 96). The purpose of the analysis was to ascertain and identify the details of each of the tasks and therein the actions, decisions, and probable errors. An extract of the analysis is presented in Table 3.5.

Table 3.4 HTA Extract for Identify Vehicle Movement and Patterns

7.3. Identify vehicle movement and patterns (Plan 7.3: Do 7.3.1. then 7.3.2. then 7.3.3.
then 7.3.4. then 7.3.5. then 7.3.6. then EXIT)

    7.3.1. Identify motor vehicle pattern (Plan 7.3.1: Do 7.3.1.1. and 7.3.1.2. in any
order then EXIT)

        7.3.1.1. Identify on-site patterns (Plan 7.3.1.1: Do 7.3.1.1.1.–7.3.1.1.7. in any order
then EXIT)

            7.3.1.1.1. Identify vehicle and traffic types

            7.3.1.1.2. Identify vehicles purposes (origins and destinations)

            7.3.1.1.3. Identify vehicle volumes

            7.3.1.1.4. Identify schedules

            7.3.1.1.5. Identify peak loads

            7.3.1.1.6. Identify parking locations

            7.3.1.1.7. Identify of access to and from the site

        7.3.1.2. Identify off-site patterns (Plan 7.3.1.2: Do 7.3.1.2.1.–7.3.1.2.7. in any order
then EXIT)

            7.3.1.2.1. Identify vehicle and traffic types

            7.3.1.2.2. Identify vehicles, purposes (origins and destinations)

            7.3.1.2.3. Identify vehicle volumes

            7.3.1.2.4. Identify schedules

            7.3.1.2.5. Identify peak loads

            7.3.1.2.6. Identify parking locations

            7.3.1.2.7. Identify of access to and from the site

    7.3.2. Identify motor vehicle pattern improvements

Table 3.5 Is an Extract of Task-Decomposition Analysis for the Task "7.3.1.1.1
Identify Vehicle and Traffic Types"

Task step description: 7.3.1.1.1 Identify vehicle and traffic types	*Task complexity:* Low
Initiating event: Identifying on-site vehicle patterns	*Task difficulty:* Low
Nature of the task: Observation	*Skills required:* Medium— equipment set up and checks
Job aids required: Video and survey charts	*Time taken:* High—36 h over 7 days
Actions required: Confirm equipment availability Equipment set up on-site 24 h recording	*Probable errors:* Equipment failure Failure to check equipment Vehicle and traffic misidentification
Decisions required: Is this the correct site? Have the data been captured?	*Error consequence:* Inaccurate vehicle identification Inaccurate on-site vehicle pattern data
Outputs from task: Vehicle and traffic type identification	*Concurrent tasks:* Identify vehicle and traffic types off-site

3.4.8 Related Methods

The task-decomposition technique relies on a number of data collection techniques for its input. The initial task description required is normally provided by conducting an HTA for the task under analysis. Data collection for the task-decomposition analysis can involve any number of HF methods, including observational techniques, interviews, walkthrough analysis, and questionnaires.

3.4.9 Approximate Training and Application Times

As a number of techniques are used within a task-decomposition analysis, the training time associated with the technique is high. Not only would an inexperienced practitioner require training in the task-decomposition technique itself (which incidentally would be minimal), but they would also require training in HTA and any techniques that would be used in the data collection part of the analysis. Moreover, due to the exhaustive nature of a task-decomposition analysis, the associated application time is also very high. Kirwan and Ainsworth (1992) suggested that task decomposition can be a lengthy process, and that its main disadvantage is the huge amount of time associated with collecting the required information.

3.4.10 Reliability and Validity

At present, no data regarding the reliability and validity of the technique are offered in the literature. It is apparent that such a technique may suffer from reliability problems, as a large proportion of the analysis is based upon the analyst's subjective judgment.

3.4.11 Tools Needed

The tools needed for a task-decomposition analysis are determined by the scope of the analysis and the techniques used for the data collection process. Task decomposition can be conducted using just pen and paper. However, it is recommended that for the data collection process, visual and audio-recording equipment should be employed. The urban system under analysis is also required in some form—either in procedurally, temporary environment or in operational form.

3.4.12 Flowchart

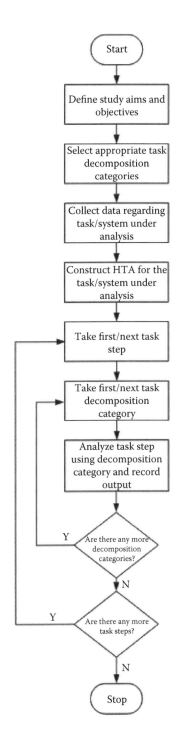

3.4.13 Recommended Text(s)

Kirwan, B., & Ainsworth, L. K. (1992). *A guide to task nalysis.* London: Taylor & Francis Group.

3.5 Operation Sequence Diagrams

3.5.1 Background and Applications

OSDs are used to graphically describe the activity and interaction between teams of agents within a system. According to Kirwan and Ainsworth (1992), the original purpose of OSD analysis was to represent complex multiperson tasks. The output of an OSD graphically depicts the task process, including the tasks performed and the interaction between operators over time, using standardized symbols. There are various forms of OSD, ranging from a simple flow diagram representing task order to more complex diagrams that account for team interaction and communication. OSDs have recently been used by the authors for the analysis of command and control in a number of domains, including the fire service, naval warfare, aviation, energy distribution, air traffic control, and rail domains.

3.5.2 Domain of Application

The technique was originally used in the nuclear power and chemical process industries. However, the technique is generic and can be applied in any domain.

3.5.3 Application in Land Use Planning and Urban Design

OSDs allow for the exploration of multistakeholder environments. This may be within the context of land use planning or design procedural tasks or indeed where there are multiple distinct users of an urban setting over time.

3.5.4 Procedure and Advice

  Step 1: Define the Task(s) Under Analysis
    The first step in an OSD analysis is to define the task(s) or scenario(s) under analysis. These should be defined clearly, including the activity and agents involved.

Step 2: Data Collection

To construct an OSD, the analyst must obtain specific data regarding the task or scenario under analysis. It is recommended that the analyst use various forms of data collection in this phase. Observational study should be used to observe the task (or similar types of task) under analysis. Interviews with personnel involved in the task (or similar tasks) should also be conducted. The type and amount of data collected in Step 2 are dependent upon the analysis requirements. The more exhaustive the analysis is intended to be, the more data collection techniques should be employed.

Step 3: Describe the Task or Scenario Using HTA

Once the data collection phase is completed, a detailed task analysis should be conducted for the scenario under analysis. The type of task analysis is determined by the analyst, and, in some cases, a task list will suffice. However, it is recommended that an HTA is conducted for the task under analysis.

Step 4: Construct the OSD Diagram

Once the task has been described adequately, the construction of the OSD can begin. The process begins with the construction of an OSD template. The template should include the title of the task or scenario under analysis, a timeline and a row for each agent involved in the task. An OSD template used during the analysis of the stakeholder interactions within the environmental assessment of a site analysis is presented in Figure 3.4. To construct the OSD, it is recommended that the analyst walks through the HTA of the task under analysis, creating the OSD simultaneously. The symbols involved in a particular task step should be linked by directional arrows to represent the flow of activity during the scenario. Each symbol in the OSD should contain the corresponding task step number from the HTA of the scenario. The artifacts used during the communications should also be annotated onto the OSD.

Step 5: Calculate Operation Loading Figures

Operation loading figures can be calculated for each stakeholder or agent involved in the scenario from the OSD.

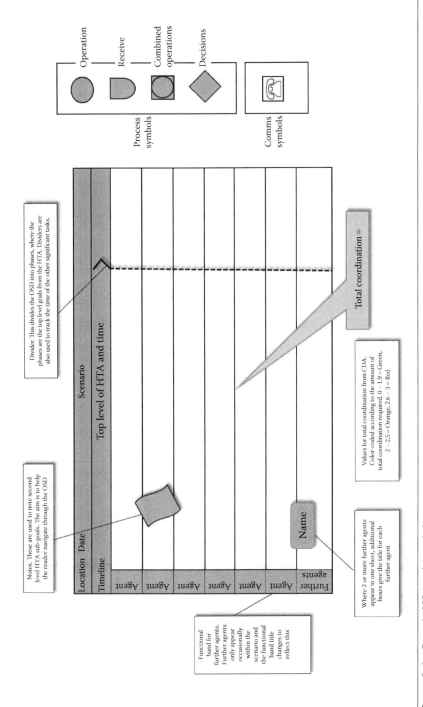

Figure 3.4 Example OSD template and glossary.

The figures are calculated for each OSD operator or symbol used, for example, operation, delay, and decision. These operation loading figures refer to the extent that each agent was involved in the operation in question during the scenario.

Step 6: Overlay Additional Analyses Results

One of the endearing features of the OSD technique is that additional analysis results can easily be added to the OSD. According to the analysis requirements, additional task features can also be annotated onto the OSD. For example, in the analysis of environmental assessment activity in a variety of domains, the authors annotated coordination values (from a coordination demands analysis: CDA) between team members for each task step onto the OSD.

3.5.5 *Advantages*

- OSDs provide an exhaustive analysis of the task in question. The flow of the task is represented in terms of activity and information, the type of activity and the stakeholders involved are specified, whereas a timeline of the activity, the communications between stakeholders involved in the task, the technology used, and also a rating of total coordination for each teamwork activity are also provided. The flexibility of the technique also permits the analyst to add further analysis outputs onto the OSD, adding to its exhaustiveness.
- They are particularly useful for analyzing and representing distributed teamwork or collaborated activity.
- They are useful for demonstrating the relationship between tasks, technology, and team members.
- They demonstrate high face validity (Kirwan and Ainsworth, 1992).
- They have been used extensively in the past and have been applied in a variety of domains.
- The OSD technique is very flexible and can be modified to suit the analysis needs.

- The WESTT software package can be used to automate a large proportion of the OSD procedure.
- Despite its exhaustive nature, the OSD technique requires only minimal training.

3.5.6 Disadvantages

- The application time for an OSD analysis is lengthy. Constructing an OSD for large, complex tasks can be extremely time-consuming, and the initial data collection stage adds further time to the analysis.
- The construction of large, complex OSDs is also quite a laborious and taxing process.
- OSDs can become cluttered and confusing (Kirwan and Ainsworth, 1992).
- Their output can become large and unwieldy.
- The present OSD symbols are limited for certain applications.
- The reliability of the technique is questionable. Different analysts may interpret the OSD symbols differently.

3.5.7 Related Methods

Various types of OSD exist, including temporal operational sequence diagrams, partitioned operational sequence diagrams, and spatial operational sequence diagrams (Kirwan and Ainsworth, 1992). During the OSD data collection phase, traditional data collection procedures such as observational study and interviews are typically employed. Task-analysis techniques such as HTA are also used to provide the input for the OSD. Timeline analysis may also be employed in order to construct an appropriate timeline for the task or scenario under analysis.

3.5.8 Approximate Training and Application Times

No data regarding the training and application time associated with the OSD technique is available in the literature. However, it is apparent that the training time for such a technique would be minimal. The application time for the technique is very high, including the initial data collection phase of interviews and observational analysis, and also the construction of an appropriate HTA for the task under analysis. The construction of the OSD in particular is a very time-consuming process. A typical OSD normally can take up to one week to construct.

3.5.9 Reliability and Validity

According to Kirwan and Ainsworth, OSD techniques possess a high degree of face validity. The intra-analyst reliability of the technique may be suspect, as different analysts may interpret the OSD symbols differently.

3.5.10 Tools Needed

When conducting an OSD analysis, pen and paper may be sufficient. However, to ensure that data collection is comprehensive, it is recommended that video- or audio-recording devices are used in conjunction with this. For the construction of the OSD, it is recommended that a suitable drawing package such as Microsoft Visio is used. The WESTT software package (Houghton et al., 2008) can also be used to automate a large portion of the OSD procedure. WESTT constructs the OSD based upon an input of observational data for the scenario under analysis.

3.5.11 Example

The following example represents the OSD for the *identify natural physical features* task of the larger Site Analysis HTA. Table 3.6 represents the HTA extract utilized for establishing the OSD, whereas Figure 3.5 is an extract from the OSD for the site-analysis scenario.

Table 3.6   Extract of HTA for Site Analysis

5. Identify natural physical features (Plan 5: Do 5.1, then 5.2 and 5.3, then EXIT)
  5.1. Establish topography (Plan 5.1: Do 5.1.1, then 5.1.2 and 5.1.3, then EXIT)
     5.1.1. (Surveyor contact State Gov.) Establish site contours
     5.1.2. (Surveyor + State Gov.) Identify major topographic features
     5.1.3. (Surveyor) Forward details/results to site analysis manager (planner)
  5.2. Identify drainage patterns (Plan 5.2: Do 5.2.1, then 5.2.2, 5.2.3, 5.2.4, respectively, then EXIT)
     5.2.1. (Hydrologist contacts State Gov.) Identify direction of surface drainage
     5.2.2. (Hydrologist) Identify arteries of water collection
     5.2.3. (Hydrologist + Local Gov.) Identify areas of flooding and inundation
     5.2.4. (Hydrologist) Forward details/results to site-analysis manager (planner)

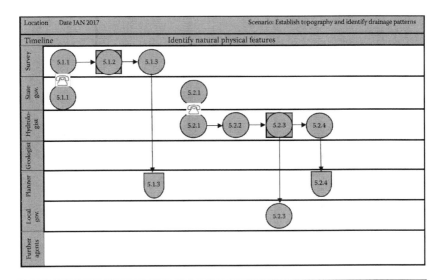

Figure 3.5   Extract from the OSD for the site-analysis scenario.

3.5.12 Flowchart

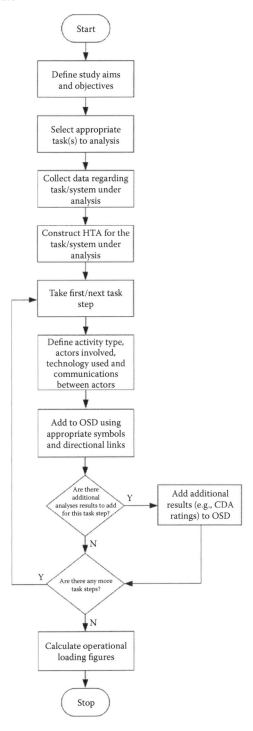

3.6 Critical Path Analysis

CPA is a popular technique in project management (Lockyer and Gordon, 1991) and is used to estimate the duration of a project in which some activities can be performed in parallel. The assumption is that a given task cannot start until all preceding tasks that contribute to it are complete. This means that some tasks might be completed, and the process is waiting for other tasks before it is possible to proceed. The tasks that are completed but are waiting for others are said to be *floating*, that is, they can shift their start times with little impact on the overall process. On the other hand, tasks that the others wait for are said to lie on the critical path, and any change to these tasks will have an impact on the overall process time. It is possible to apply these ideas to any time-based activity, including human performance.

To calculate CPA, it is necessary to know the order in which tasks are performed, their duration, and their dependency. The notion of dependency is, for traditional CPA, based on the question of what tasks need to be completed before another task is allowed to commence. However, when applied to human performance models, dependency offers a richer conceptual framework in that it allows consideration of parallel activity. Models based on CPA can be constructed to represent some aspects of parallel activity, which can provide more accurate estimates of performance time (Baber and Mellor, 2001; Gray et al., 1993).

3.6.1 Describing Dependency

To introduce the concept of dependency, it is necessary to make assumptions about the order in which tasks are performed and the nature of the tasks themselves. Clearly, some tasks need to be completed before others can start (which is central to traditional CPA modeling). This means that we can consider temporal dependency as the first stage in constructing a CPA model. However, temporal dependency tells us nothing about why some tasks can be performed in parallel. To consider this issue, we turn to notions of multiple resources.

3.6.2 Multiple Resources

For the HF community, it is convenient to assume that tasks involving different modalities, such as speaking and looking, can be performed with little interference. This assumption is not without criticism, and there are several experiments that we will not consider here which suggest that interference can occur at the stage of central processing of information. This means that, like many assumptions within HF, what serves as a useful aid in engineering applications is not necessarily supported as a generalizable component of human cognition. Wickens (1992) amalgamated a considerable amount of research on multiple task performance to propose a theory of multiple attentional resources. The theory proposes a general pool (or reservoir) of attentional resources that is shared across stages of human information processing: as the demands of one stage increase, so the resources available to other stages diminish. To manage this distribution of resource, the theory assumes that there are two subpools: one for visuospatial resources and one for verbal-acoustic resources. Such a model would help to determine the possibility of tasks being performed in series or parallel, that is, two *visual* tasks would need to be performed in series (for the simple reason that one cannot look in two places at the same time), but an *auditory* and *visual* task could possibly be performed in parallel, for example, the (visual) monitoring of displays could be performed in parallel with the (auditory) hearing of an alarm. The suggestion is that, as tasks draw from the same subpool, their interference requires serial processing, but if they use different subpools, they can be performed in parallel. A complication with this assumption is that the various stages of processing might draw on different versions of the subpool, for example, at the input stage, the *subpool* could be constrained by sensory limitations (e.g., you cannot look at two places at once but need to move your eyes between the places), and at the output stage, the *subpool* would be constrained by response mechanisms, for example, speaking or pressing buttons. Thus, at the observable stages of human response, it is possible to make certain assumptions relating to the manner in which information is presented to the person or responses are made. However, the central processing stage is not so amenable to reductionism, and it is not entirely clear what *codes* are used to represent information. Although this could be a problem for experimental psychology, HF tends to stick with the observable aspects of input/output and uses these notions for characterizing tasks. So, we would consider *input* in terms of vision or hearing and *output* in terms of speech or manual response (left or right hand). For the purposes of this approach, we also include a generalized *cognition* component (it would be possible to assume that cognition is performed using different codes and to include some additional components, but this is neither substantiated by research nor particularly necessary). The list of codes used in this analysis is as follows:

- Visual
- Auditory
- Spoken response
- Manual response (left)
- Manual response (right)
- Cognition

3.6.3 Domain of Application

Primarily HCI, but also generic.

3.6.4 Application in Land Use Planning and Urban Design

Due to the close association with project management, many in the LUP & UD disciplines will be familiar with CPA. It is an inherently useful approach to better understanding the requirements and resources for projects and tasks within LUP & UD.

3.6.5 Procedure and Advice

In this chapter, construction of a CPA model is based upon a method initially developed by Gray et al. (1993) and further refined by Baber and Mellor (2001). The method may be proceduralized as follows.

Step 1: Analyze the Tasks to be Modeled

The tasks need to be analyzed in fine detail if they are to be modeled by multimodal CPA. HTA can be used, but it needs to be conducted down to the level of individual task units. This fine-grain level of analysis is essential if reasonable predictions of response times are to be made. Figure 3.6 is the HTA of a generic development assessment process with an information request and decision. The full HTA Plan for the

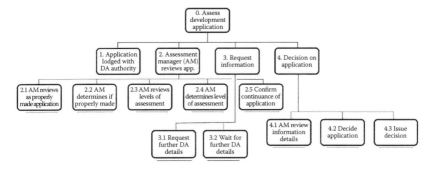

Figure 3.6 HTA code assessable development application with information request.

task Assess Development Application can be found at the end of this chapter in Table 3.10.

Step 2: Order the Tasks

This requires an initial sketch (drawn as a flowchart) of the task sequences in terms of temporal dependency. At this stage, the analyst is considering whether more than one task might feed into subsequent tasks. Figure 3.7 represents the sequences of tasks over time associated with a code assessable development application process, including a request for information, and the issue of a decision.

Step 3: Allocate Subtasks to Modality

Each unit task then needs to be assigned to a modality. For the purposes of land use planning and urban design tasks, these modalities may be as follows:

- *Visual tasks*: For example, looking at plans, maps, displays, or written notes and procedures.
- *Communication*: Sending and requesting written and mapping information and providing decisions in writing.

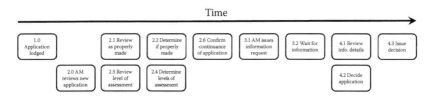

Figure 3.7 Development assessment tasks on temporal dependency.

- *Cognition*: For example, making decisions about whether something is properly filled out or complaint; or if to intervene and selecting intervention strategies.
- *Manual tasks*: For example, taking notes, collecting data, recording details, drawing and presenting, and site visits.
- *Speech tasks*: Providing and receiving verbal advice, or talking and consulting with colleagues or the community. Table 3.7 shows each task associated with its appropriate modularity.

Step 4: Sequence the Subtasks in a Multimodal CPA Diagram

The tasks are put into the order of occurrence, checking the logic for parallel and serial tasks. For serial tasks, the logical sequence is determined by the task analysis. For parallel tasks, the modality determines their placement in the representation.

Step 5: Allocate Timings to the Subtasks

Timings for the tasks are derived from a number of sources. For the purposes of this exercise, the timings used are based on the number of business days associated with the development assessment process in Queensland, Australia.

Step 6: Determine the Time to Perform the Whole Task

The time in which the task may be performed can be found by tracing through the CPA using the longest node-to-node

Table 3.7   Development Assessment Tasks Organized in Modularity

VISUAL TASKS	COMMUNICATION	COGNITION	MANUAL TASK	SPEECH TASK
Assessment manager reviews application	Request further development application (DA) details	Determine if properly made		
Review as properly made application	Provide confirmation of rejection of application	Determine levels of assessment		
Review levels of assessment	Provide confirmation of continuance of application	Decide application		
Review requested information	Issue decision			

values. The calculations in CPA are fairly simple, providing two basic rules are followed:

1. On the *forward-pass*, take the longest time.
2. On the *backward-pass*, take the shortest time.

The calculation can be most easily represented in the form of a diagram representing the tasks and their start/finish times. As Figure 3.8 illustrates, each task is represented as a box containing its number and name, its duration, the earliest start time (EST), and latest finish time (LFT) and float.

Having established a sequence (based on temporal and modality dependency) and associated tasks with times, the final stage is to perform the calculation. In this section, the boxes defined earlier are presented in Figure 3.9 and Table 3.8.

To undertake the calculations

1. Begin with an EST on 0 for the first activity.
2. Calculate the EFT as the sum of the EST and duration.
3. Use the EFT for one task as the EST of the next task (unless there is a choice of EFTs, in which case take the largest).
4. Continue calculating the EFT until the end.
5. Set the LFT to equal the EFT of the final task.
6. Subtract duration from LFT to get LST.
7. Insert LST as LFT on previous task (unless there is a choice, in which case take the smallest).
8. Continue until first task reached.

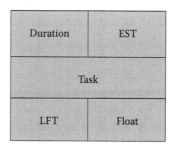

Figure 3.8 Key for each node in CPA diagram.

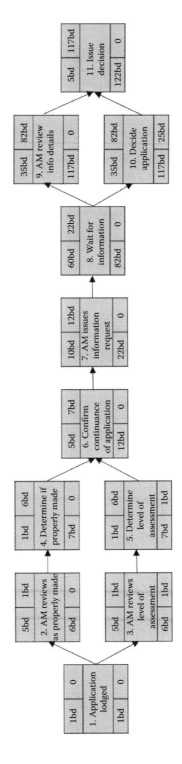

Figure 3.9   Summary analysis.

Table 3.8 CPA Calculations for Development Assessment Example

TASK	DURATION (BUSINESS DAYS)	EST	EFT	LST	LFT	FLOAT
1. Application lodged	1bd	0	1bd	0bd	1bd	0
2. AM reviews as properly made	5bd	1bd	6bd	1bd	6bd	0
3. AM reviews level of assessment	5bd	1bd	6bd	1bd	6bd	1bd
4. Determine if properly made	1bd	6bd	7bd	6bd	7bd	0
5. Determine level of assessment	1bd	6bd	7bd	6bd	7bd	1bd
6. Confirm continuance of application	5bd	7bd	12bd	7bd	12bd	0
7. AM issues information request	10bd	12bd	22bd	12bd	22bd	0
8. Wait for information	60bd	22bd	82bd	22bd	82bd	0
9. AM review info details	35bd	82bd	117bd	82bd	117bd	0
10. Decide application	35bd	82bd	117bd	82bd	117bd	25bd
11. Issue decision	5bd	117bd	122bd	117bd	122bd	0

3.6.6 *Advantages*

- CPA allows the analyst to gain a better understanding of the task via splitting the task into the activities that need to be carried out to ensure successful task completion.
- It allows the consideration of parallel unit task activity (Baber and Mellor, 2001).
- It gives predicted performance task times for the full task and also for each task step.
- It determines a logical, temporal description of the task in question.
- It does not require a great deal of training.
- It is a structured and comprehensive procedure.
- It can accommodate parallelism in user performance.
- It provides reasonable fit with observed data.

3.6.7 Disadvantages

- CPA can be tedious and time-consuming for complex tasks.
- It only models error-free performance and cannot deal with unpredictable events such as those seen in human–machine interactions.
- Its modality can be difficult to define.
- It can only be used for activities that can be described in terms of performance times.
- Times are not available for all actions.
- It can be overly reductionistic, particularly for tasks that are mainly cognitive in nature.

3.6.8 Related Methods

CPA is one of a number of performance-time prediction methods that also include the keystroke-level model (KLM) (Card et al., 1983) and timeline analysis.

3.6.9 Approximate Training and Application Times

Although no data regarding the training and application time of CPA are available, it is suggested that both the training time the application time would be low, although this is dependent upon the task under analysis. For complex, larger tasks, the application time would be high.

In a review of interface methods, Stanton et al. (2014) proposed that CPA was a time-consuming technique compared with heuristic analysis, layout analysis, and HTA. They estimated that to analyze multimodal in-vehicle car interfaces, 2–4 h are required to collect the data and 8–10 h are needed to analyze them.

3.6.10 Reliability and Validity

Baber and Mellor (2001) compared predictions using CPA with the results obtained from user trials and found that the *fit* between observed and predicted values had an error of less than 20 percent. This suggests that the approach can provide robust and useful approximations of human performance.

3.6.11 Tools Needed

CPA can be conducted using pen and paper.

3.6.12 Flowchart

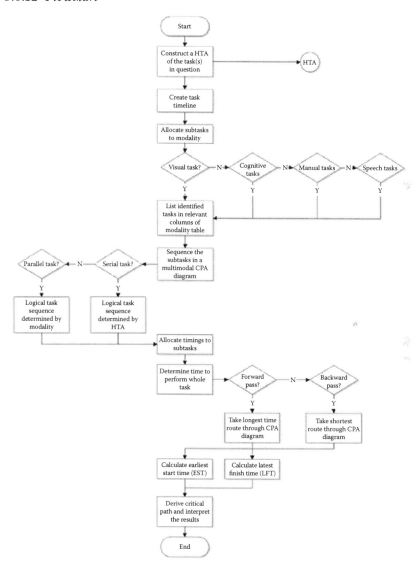

Table 3.9   Full HTA Plan for Task Conduct Site Analysis

---

0. Conduct site analysis
Plan 0: Do 1, then 2, 3, 4, 5, and 6 in any order, then 7, 8, 9, 10, and 11 in any order, then EXIT

1. Identify legal descriptions (Plan 1: Do 1.1, then 1.2, 1.3, and 1.4, then EXIT)
    1.1. Identify the legal description of the property
    1.2. Identify legal stakeholders (Plan 1.2: Do 1.2.1, then 1.2.2, then EXIT)
        1.2.1. Identify name of property owner
        1.2.2. Identify government departments with property jurisdictions
    1.3. Identify property covenants and legal restrictions
    1.4. Identify any projected or proposed changes to legal descriptions
2. Analyze Location (Plan 2: Do 2.1, then 2.2 and 2.3, respectively, then EXIT)
    2.1. Identify location (Plan 2.1: Do 2.1.1, then 2.1.2 and 2.1.3, respectively, then EXIT)
        2.1.1. Identify the city in the state
        2.1.2. Identify the neighborhood in the city
        2.1.3. Identify the site in the neighborhood
    2.2. Establish distances (Plan 2.2: Do 2.2.1–2.2.5 in any order, then EXIT)
        2.2.1. Establish the distance between the site and the city commercial functions
        2.2.2. Establish the distance between the site and the city retail functions
        2.2.3. Establish the distance between the site and the city residential functions
        2.2.4. Establish the distance between the site and the city community functions
        2.2.5. Establish the distance between the site and the city industrial functions
    2.3. Establish travel times (Plan 2.3: Do 2.3.1–2.3.5 in any order, then EXIT)
        2.3.1. Est. travel times between the site and the city commercial functions (Plan 2.3.1: Do 2.3.1.1–2.3.1.4 in any order, then EXIT)
            2.3.1.1. Est. walking travel time
            2.3.1.2. Est. cycling travel time
            2.3.1.3. Est. public transport travel time
            2.3.1.4. Est. private vehicle travel time
        2.3.2. Est. travel times between the site and the city retail functions (Plan 2.3.2: Do 2.3.2.1–2.3.2.4 in any order, then EXIT)
            2.3.2.1. Est. walking travel time
            2.3.2.2. Est. cycling travel time
            2.3.2.3. Est. public transport travel time
            2.3.2.4. Est. private vehicle travel time
        2.3.3. Est. travel times between the site and the city residential functions (Plan 2.3.3: Do 2.3.3.1–2.3.3.4 in any order, then EXIT)
            2.3.3.1. Est. walking travel time
            2.3.3.2. Est. cycling travel time
            2.3.3.3. Est. public transport travel time
            2.3.3.4. Est. private vehicle travel time
        2.3.4. Est. travel times between the site and the city community functions (Plan 2.3.4: Do 2.3.4.1–2.3.4.4 in any order, then EXIT)
            2.3.4.1. Est. walking travel time
            2.3.4.2. Est. cycling travel time
            2.3.4.3. Est. public transport travel time
            2.3.4.4. Est. private vehicle travel time

(Continued)

Table 3.9 (*Continued*)   Full HTA Plan for Task Conduct Site Analysis

---

      2.3.5. Est. travel times between the site and the city industrial functions (Plan 2.3.5: Do 2.3.5.1–2.3.5.4 in any order, then EXIT)

         2.3.5.1. Est. walking travel time

         2.3.5.2. Est. cycling travel time

         2.3.5.3. Est. public transport travel time

         2.3.5.4. Est. private vehicle travel time

3. Identify size and planning controls (Plan 3: Do 3.1, then 3.2, then EXIT)

   3.1. Establish dimensions (Plan 3.1: Do 3.1.1, then 3.1.2 and 3.1.3 in any order, then EXIT)

      3.1.1. Survey the boundaries of the site

      3.1.2. Establish dimensions of any street rights of way of the site

      3.1.3. Establish dimensions of any site easements

   3.2. Identify planning controls (Plan 3.2: Do 3.2.1, then 3.2.2 and 3.2.3 in any order, then 3.2.4 and 3.2.5, then EXIT)

      3.2.1. Identify present site zoning (Plan 3.2.1: Do 3.2.1.1, then 3.2.1.2, 3.2.1.3, and 3.2.1.4, then EXIT)

         3.2.1.1. Identify site zoning

         3.2.1.2. Identify building height restrictions

         3.2.1.3. Establish m2 and m3 of buildable area

         3.2.1.4. Establish site car-parking requirements

      3.2.2. Identify any local area plans

      3.2.3. Identify other planning controls

      3.2.4. Establish proposed zoning (Plan 3.2.4: Do 3.2.4.1, then 3.2.4.2, 3.2.4.3, and 3.2.4.4, respectively, then EXIT)

         3.2.4.1. Identify present site zoning

         3.2.4.2. Identify building height restrictions

         3.2.4.3. Establish m2 and m3 of buildable area

         3.2.4.4. Establish site car-parking requirements

      3.2.5. Identify any proposed changes to planning controls

4. Analyze Neighborhood (Plan 4: Do 4.1, then 4.2 and 4.3, respectively, then EXIT)

   4.1. Establish locations of neighborhood functions (Plan 4.1: Do 4.1.1–4.2.4 in any order, then EXIT)

      4.1.1. Identify commercial functions

      4.1.2. Identify retail functions

      4.1.3. Identify residential functions

      4.1.4. Identify community functions

   4.2. Establish neighborhood context (Plan 4.2: Do 4.2.1, then 4.2.2, 4.2.3, 4.2.4, and 4.1.5, respectively, then EXIT)

      4.2.1. Identify neighborhood urban form (Plan 4.2.1: Do 4.2.1.1–4.2.1.4 in any order, then EXIT)

         4.2.1.1. Identify patterns

         4.2.1.2. Identify layout

         4.2.1.3. Identify density

         4.2.1.4. Identify structures

(Continued)

Table 3.9 (*Continued*)   Full HTA Plan for Task Conduct Site Analysis

4.2.2. Identify neighborhood open spaces (Plan 4.2.2: Do 4.2.2.1 then 4.2.2.2, then EXIT)
 4.2.2.1. Identify existing public and private open space
 4.2.2.2. Identify proposed public and private open space
4.2.3. Identify neighborhood built form (Plan 4.2.3: Do 4.2.3.1 then 4.2.3.2, then EXIT)
 4.2.3.1. Identify existing built form (Plan 4.2.3.1: Do 4.2.3.1.1 then 4.2.3.1.2 then
   4.2.3.1.3, then EXIT)
 4.2.3.1.1. Identify building ages
 4.2.3.1.2. Identify building conditions
 4.2.3.1.3. Identify architectural patterns (Plan 4.2.3.1.3:
   Do 4.2.3.1.3.1–4.2.3.1.3.5 in any order, then EXIT)
   4.2.3.1.3.1. Identify roof forms
   4.2.3.1.3.2. Identify materials
   4.2.3.1.3.3. Identify color
   4.2.3.1.3.4. Identify heights and bulk
   4.2.3.1.3.5. Identify articulations
4.2.4. Identify neighborhood microclimate (Plan 4.2.4: Do 4.2.4.1 and 4.2.4.2 in any order
 then do 4.2.4.3, then EXIT)
 4.2.4.1. Identify solar orientation and access
 4.2.4.2. Identify breezes and wind direction
 4.2.4.3. Identify seasonal variations in sun and wind
4.2.5. Identify any locations of neighborhood significance (Plan 4.2.5: Do 4.2.5.1–4.2.5.4
 in any order, then EXIT)
 4.2.5.1. Identify historic significance
 4.2.5.2. Identify views of significance
 4.2.5.3. Identify situations of significance
 4.2.5.4. Identify districts of significance
4.3. Establish travel times between site and neighborhood functions (Plan 4.3: Do 4.3.1–4.3.4
 in any order, then EXIT)
 4.3.1. Identify travel times to commercial functions (Plan 4.3.1: Do 4.3.1.1–4.3.1.3 in any
   order, then EXIT)
   4.3.1.1. Est. walking travel time
   4.3.1.2. Est. cycling travel time
   4.3.1.3. Est. public transit travel time
 4.3.2. Identify travel times to retail functions (Plan 4.3.2: Do 4.3.2.1–4.3.2.3 in any order,
   then EXIT)
   4.3.2.1. Est. walking travel time
   4.3.2.2. Est. cycling travel time
   4.3.2.3. Est. public transit travel time
 4.3.3. Identify travel times to residential functions (Plan 4.3.3: Do 4.3.3.1–4.3.3.3 in any
   order, then EXIT)
   4.3.3.1. Est. walking travel time
   4.3.3.2. Est. cycling travel time
   4.3.3.3. Est. public transit travel time

(*Continued*)

Table 3.9 (*Continued*) Full HTA Plan for Task Conduct Site Analysis

---

4.3.4. Identify travel times to community functions (Plan 4.3.4: Do 4.3.4.1–4.3.4.3 in any order, then EXIT)

    4.3.4.1. Est. walking travel time

    4.3.4.2. Est. cycling travel time

    4.3.4.3. Est. public transit travel time

5. Identify natural physical features (Plan 5: Do 5.1, then 5.2 and 5.3, respectively, then EXIT)

    5.1. Establish topography (Plan 5.1: Do 5.1.1 then 5.1.2, then EXIT)

        5.1.1. Establish site contours

        5.1.2. Identify major topographic features (Plan 5.1.2: Do 5.1.2.1–5.1.2.6 in any order, then EXIT)

            5.1.2.1. Identify slopes

            5.1.2.2. Identify flay areas

            5.1.2.3. Identify high points

            5.1.2.4. Identify low points

            5.1.2.5. Identify ridges

            5.1.2.6. Identify valleys

    5.2. Identify drainage patterns (Plan 5.2: Do 5.2.1, then 5.2.2 and 5.2.3, respectively, then EXIT)

        5.2.1. Identify direction of surface drainage (Plan 5.2.1: Do 5.2.1.1 and 5.2.1.2 in any order, then EXIT)

            5.2.1.1. Identify drainage on to the site

            5.2.1.2. Identify drainage off the site

        5.2.2. Identify arteries of water collection (Plan 5.2.: Do 5.2.2.1 and 5.2.2.2 in any order, then EXIT)

            5.2.2.1. Identify major arteries

            5.2.2.2. Identify minor arteries

        5.2.3. Identify areas of flooding and inundation (Plan 5.2.3: Do 5.2.3.1 and 5.2.3.1 in any order, then EXIT)

            5.2.3.1. Identify flooding on-site

            5.2.3.2. Identify flooding off-site

    5.3. Manage biophysical features (Plan 5.3: Do 5.3.1–5.3.4 in any order, then EXIT)

        5.3.1. Determine site flora value (Plan 5.3.1: Do 5.3.1.1 then 5.3.1.2, then EXIT)

            5.3.1.1. Assess flora

            5.3.1.2. Decide flora value (Plan 5.3.1.2: Do 5.3.1.2.1–5.3.1.2.5 as required based on assessment, then EXIT)

                5.3.1.2.1. Select preservation

                5.3.1.2.2. Select protection

                5.3.1.2.3. Select reinforcement

                5.3.1.2.4. Select alteration

                5.3.1.2.5. Select removal

        5.3.2. Determine site fauna value (Plan 5.3.2: Do 5.3.2.1 then 5.3.2.2, then EXIT)

            5.3.2.1. Assess fauna

            5.3.2.2. Decide fauna value (Plan 5.3.2.2: Do 5.3.2.2.1–5.3.2.2.5 as required based on assessment, then EXIT)

(Continued)

Table 3.9 (*Continued*)   Full HTA Plan for Task Conduct Site Analysis

5.3.2.2.1. Select preservation

5.3.2.2.2. Select protection

5.3.2.2.3. Select reinforcement

5.3.2.2.4. Select alteration

5.3.2.2.5. Select removal

5.3.3. Determine geology outcrop value (Plan 5.3.3: Do 5.3.3.1 then 5.3.3.2, then EXIT)

5.3.3.1. Assess geology outcrops

5.3.3.2. Decide geology outcrop value (Plan 5.3.3.2: Do 5.3.3.2.1–5.3.3.2.5 as required based on assessment, then EXIT)

5.3.3.2.1. Select preservation

5.3.3.2.2. Select protection

5.3.3.2.3. Select reinforcement

5.3.3.2.4. Select alteration

5.3.3.2.5. Select removal

5.3.4. Identify soil type and bearing capacity (Plan 5.3.4: Do 5.3.4.1 then 5.3.4.2, then EXIT)

5.3.4.1. Identify soil type distribution over the site

5.3.4.2. Identify soil type at different sublevels

6. Identify human-made features (Plan 6: Do 6.1, then 6.2 and 6.3, followed by 6.4, then EXIT)

6.1. Identify site buildings (Plan 6.1: Do 6.1.1, then 6.2.2, 6.3.2, 6.3.4–6.3.8 in any order, then EXIT)

6.1.1. Establish location

6.1.2. Establish significance on registry listing

6.1.3. Establish structural integrity

6.1.4. Establish size and shape

6.1.5. Establish height

6.1.6. Identify character and materials

6.1.7. Identify interior layout

6.1.8. Establish architectural qualities

6.2. Identify external open spaces on-site (Plan 6.2: Do 6.2.1, then 6.2.2, 6.2.3, 6.2.3, respectively, then EXIT)

6.2.1. Establish location

6.2.2. Establish size and shape

6.2.3. Identify character and materials

6.2.4. Establish structural and surface integrity

6.3. Identify walls and fences on-site (Plan 6.3: Do 6.3.1, then 6.3.2, 6.3.3, 6.3.4, respectively, then EXIT)

6.3.1. Establish location

6.3.2. Identify character and materials

6.3.3. Establish structural and surface integrity

6.3.4. Establish height and length

6.4. Identify surrounding off-site human made features (Plan 6.4: Do 6.4.1, then 6.4.2 and 6.4.3, respectively, then EXIT)

6.4.1. Identify surrounding buildings (Plan 6.4.1: Do 6.4.1.1, then 6.4.1.2, 6.4.1.3–6.4.1.6 in any order, then EXIT)

(Continued)

Table 3.9 (*Continued*)   Full HTA Plan for Task Conduct Site Analysis

6.4.1.1. Establish location

6.4.1.2. Establish significance of registry listing

6.4.1.3. Establish size and shape

6.4.1.4. Establish height

6.4.1.5. Identify character and materials

6.4.1.6. Establish architectural qualities

6.4.2. Identify surrounding walls and fences (Plan 6.4.2: Do 6.4.2.1, then 6.4.2.2 and 6.4.2.3, then EXIT)

6.4.2.1. Establish location

6.4.2.2. Identify character and material

6.4.2.3. Establish length and height

6.4.3. Identify surrounding external open spaces (Plan 6.4.3: Do 6.4.3.1, then 6.4.3.2 and 6.4.3.3, then EXIT)

6.4.3.1. Establish location

6.4.3.2. Establish size and shape

6.4.3.3. Identify character and materials

7. Determine circulation patterns (Plan 7: Do 7.1, then 7.2 and 7.3 in any order, then 7. 4, then EXIT)

7.1. Identify site and neighborhood road hierarchy

7.2. Identify active transport movement and patterns (Plan 7.2: Do 7.2.1, then 7.2.2, 7.2.3, and 7.2.4, respectively, then EXIT)

7.2.1. Identify pedestrian pathways and pattern (Plan 7.2.1: Do 7.2.1.1 and 7.2.1.2 in any order, then EXIT)

7.2.1.1. Identify on-site pathways and patterns (Plan 7.2.1.1: Do 7.2.1.1.1–7.2.1.1.5 in any order, then EXIT)

7.2.1.1.1. Identify ped. types

7.2.1.1.2. Identify ped. purposes

7.2.1.1.3. Identify ped. schedules

7.2.1.1.4. Identify ped. volumes

7.2.1.1.5. Identify locations of ped. access to and from the site

7.2.1.2. Identify off-site pathways and pattern (Plan 7.2.1.2: Do 7.2.1.2.1–7.2.1.2.5 in any order, then EXIT)

7.2.1.2.1. Identify ped. types

7.2.1.2.2. Identify ped. purposes

7.2.1.2.3. Identify ped. schedules

7.2.1.2.4. Identify ped. volumes

7.2.1.2.5. Identify locations of ped. access to and from the site

7.2.2. Identify pedestrian pattern improvements

7.2.3. Identify cyclist pathways and patterns (Plan 7.2.3: Do 7.2.3.1 and 7.2.3.2 in any order, then EXIT)

7.2.3.1. Identify on-site pathways and patterns (Plan 7.2.3.1: Do 7.2.3.1.1–7.2.3.1.5 in any order, then EXIT)

(*Continued*)

Table 3.9 (*Continued*)   Full HTA Plan for Task Conduct Site Analysis

---

        7.2.3.1.1. Identify cyclist types

        7.2.3.1.2. Identify cyclist purposes

        7.2.3.1.3. Identify cyclist schedules

        7.2.3.1.4. Identify cyclist volumes

        7.2.3.1.5. Identify locations of cyclist access to and from the site

     7.2.3.2. Identify off-site pathways and patterns (Plan 7.2.3.2: Do 7.2.3.2.1–7.2.3.2.5 in any order, then EXIT)

        7.2.3.2.1. Identify cyclist types

        7.2.3.2.2. Identify cyclist purposes

        7.2.3.2.3. Identify cyclist schedules

        7.2.3.2.4. Identify cyclist volumes

        7.2.3.2.5. Identify locations of cyclist access to and from the site

  7.2.4. Identify cyclist pattern improvements

7.3. Identify vehicle movement and patterns (Plan 7.3: Do 7.3.1, then 7.3.2, 7.3.3, 7.3.4, 7.3.5, and 7.3.6, respectively, then EXIT)

  7.3.1. Identify motor vehicle pattern (Plan 7.3.1: Do 7.3.1.1 and 7.3.1.2 in any order, then EXIT)

     7.3.1.1. Identify on-site patterns (Plan 7.3.1.1: Do 7.3.1.1.1–7.3.1.1.7 in any order, then EXIT)

        7.3.1.1.1. Identify vehicle and traffic types

        7.3.1.1.2. Identify vehicles, purposes (origins and destinations)

        7.3.1.1.3. Identify vehicle volumes

        7.3.1.1.4. Identify schedules

        7.3.1.1.5. Identify peak loads

        7.3.1.1.6. Identify parking locations

        7.3.1.1.7. Identify access to and from the site

     7.3.1.2. Identify off-site patterns (Plan 7.3.1.2: Do 7.3.1.2.1–7.3.1.2.7 in any order, then EXIT)

        7.3.1.2.1. Identify vehicle and traffic types

        7.3.1.2.2. Identify vehicles purposes (origins and destinations)

        7.3.1.2.3. Identify vehicle volumes

        7.3.1.2.4. Identify schedules

        7.3.1.2.5. Identify peak loads

        7.3.1.2.6. Identify parking locations

        7.3.1.2.7. Identify access to and from the site

  7.3.2. Identify motor vehicle pattern improvements

  7.3.3. Identify service vehicle patterns (Plan 7.3.3: Do 7.3.3.1 and 7.3.3.2 in any order, then EXIT)

     7.3.3.1. Identify on-site patterns (Plan 7.3.3.1: Do 7.3.3.1.1–7.3.3.1.7 in any order, then EXIT)

        7.3.3.1.1. Identify vehicle and traffic types

        7.3.3.1.2. Identify vehicles purposes (origins and destinations)

        7.3.3.1.3. Identify vehicle volumes

(Continued)

Table 3.9 (*Continued*)   Full HTA Plan for Task Conduct Site Analysis

7.3.3.1.4. Identify schedules

7.3.3.1.5. Identify peak loads

7.3.3.1.6. Identify parking locations

7.3.3.1.7. Identify of access to and from the site

7.3.3.2. Identify off-site patterns (Plan 7.3.3.2: Do 7.3.3.2.1–7.3.3.2.7 in any order, then EXIT)

7.3.3.2.1. Identify vehicle and traffic types

7.3.3.2.2. Identify vehicles purposes (origins and destinations)

7.3.3.2.3. Identify vehicle volumes

7.3.3.2.4. Identify schedules

7.3.3.2.5. Identify peak loads

7.3.3.2.6. Identify parking locations

7.3.3.2.7. Identify of access to and from the site

7.3.4. Identify service vehicle pattern improvements

7.3.5. Identify public transport (also inc. taxi/uber) patterns (Plan 7.3.5: Do 7.3.5.1 and 7.3.5.2 in any order, then EXIT)

7.3.5.1. Identify on-site patterns (Plan 7.3.5.1: Do 7.3.5.1.1–7.3.5.1.7 in any order, then EXIT)

7.3.5.1.1. Identify vehicle and traffic types

7.3.5.1.2. Identify vehicles purposes (origins and destinations)

7.3.5.1.3. Identify vehicle volumes

7.3.5.1.4. Identify schedules

7.3.5.1.5. Identify peak loads

7.3.5.1.6. Identify stop (pick up/drop off) locations

7.3.5.1.7. Identify of access to and from the site

7.3.5.2. Identify off-site patterns (Plan 7.3.5.2: Do 7.3.5.2.1–7.3.5.2.7 in any order, then EXIT)

7.3.5.2.1. Identify vehicle and traffic types

7.3.5.2.2. Identify vehicles purposes (origins and destinations)

7.3.5.2.3. Identify vehicle volumes

7.3.5.2.4. Identify schedules

7.3.5.2.5. Identify peak loads

7.3.5.2.6. Identify stop (pick up/drop off) locations

7.3.5.2.7. Identify access to and from the site

7.3.6. Identify public transport pattern improvements

7.4. Identify site and neighborhood traffic generators (Plan 7.4: Do 7.4.1–7.4.3 in any order, then EXIT)

7.4.1. Identify buildings that are significant origins and destinations

7.4.2. Identify land uses that are significant origins and destinations

7.4.3. Identify events that are significant origins and destinations

8. Identify utility infrastructures (Plan 8: Do 8.1, then 8.2 and 8.3, respectively, then EXIT)

8.1. Identify the location of utilities on-site and adjacent to the site (Plan 8.1: Do 8.1.1–8.1.5 in any order, then EXIT)

(*Continued*)

Table 3.9 (*Continued*)   Full HTA Plan for Task Conduct Site Analysis

8.1.1. Identify electricity infrastructure

8.1.2. Identify gas infrastructure

8.1.3. Identify data and communication infrastructure

8.1.4. Identify water infrastructure

8.1.5. Identify sewerage infrastructure

8.2. Identify conveyance and capacity of utilities on-site and adjacent to the site (Plan 8.2: Do 8.2.1–8.2.5 in any order, then EXIT)

8.2.1. Identify electricity infrastructure conveyance and capacity

8.2.2. Identify gas infrastructure conveyance and capacity

8.2.3. Identify data and communication infrastructure conveyance and capacity

8.2.4. Identify water infrastructure conveyance and capacity

8.2.5. Identify sewerage infrastructure conveyance and capacity

8.3. Identify opportunities to connect utilities infrastructure to the site

9. Identify sensory elements (Plan 9: Do 9.1, then 9.2 and 9.3, respectively, then 9.4 and 9.5 in any order, then EXIT)

9.1. Identify views on the site from within the site (Plan 9.1: Do 9.1.1, then 9.1.2 and 9.1.3, respectively, then EXIT)

9.1.1. Assess the views (Plan 9.1.1: Do 9.1.1.1 or 9.1.1.2 as required based on assessment, then EXIT)

9.1.1.1. Select positive view

9.1.1.2. Select negative view

9.1.2. Identify if the view changes over time

9.1.3. Establish the likelihood of view conveyance

9.2. Identify views from the site to the surrounds (Plan 9.2: Do 9.2.1, then 9.2.2 and 9.2.3, respectively, then EXIT)

9.2.1. Assess the views (Plan 9.2.1: Do 9.2.1.1 or 9.2.1.2 as required based on assessment, then EXIT)

9.2.1.1. Select positive view

9.2.1.2. Select negative view

9.2.2. Identify if the view changes over time

9.2.3. Establish the likelihood of view conveyance

9.3. Identify views to the site from the surrounds (Plan 9.3: Do 9.3.1, then 9.3.2 and 9.3.3, respectively, then EXIT)

9.3.1. Assess the views (Plan 9.3.1: Do 9.3.1.1 or 9.3.1.2 as required based on assessment, then EXIT)

9.3.1.1. Select positive view

9.3.1.2. Select negative view

9.3.2. Identify if the view changes over time

9.3.3. Establish the likelihood of view conveyance

9.4. Identify significant sources of noise on and surrounding the site (Plan 9.4: Do 9.4.1, then 9.4.2 and 9.4.3, respectively, then EXIT)

9.4.1. Assess the noise (Plan 9.4.1: Do 9.4.1.1 or 9.4.1.2 as required based on assessment then EXIT)

(Continued)

Table 3.9 (*Continued*)   Full HTA Plan for Task Conduct Site Analysis

9.4.1.1. Select positive noise

9.4.1.2. Select negative noise

9.4.2. Does the noise change over time

9.4.3. Will the noise continue in the long term

9.5. Identify significant odors, smoke, or airborne pollution on and around the site (Plan 9.5: Do 9.5.1, then 9.5.2 and 9.5.3, respectively, then EXIT)

9.5.1. Assess the emission (Plan 9.5.1: Do 9.5.1.1 or 9.5.1.2 as required based on assessment, then EXIT)

9.5.1.1. Select minor

9.5.1.2. Select major

9.5.2. Does the emission change over time

9.5.3. Will the emission continue in the long term

10. Identify human and cultural elements (Plan 10: Do 10.1, then 10.2 and 10.3, respectively, then EXIT)

10.1. Document neighborhood characteristics (Plan 10.1: Do 10.1.1–10.1.4 in any order, then EXIT)

10.1.1. Identify cultural aspects (Plan 10.1.1: Do 10.1.1.1 and 10.1.1.2 in any order)

10.1.1.1. Document historical aspects

10.1.1.2. Document current aspects

10.1.2. Identify psychological aspects (Plan 10.1.2: Do 10.1.2.1 and 10.1.2.2 in any order, then EXIT)

10.1.2.1. Document historical aspects

10.1.2.2. Document current aspects

10.1.3. Identify behavioral aspects (Plan 10.1.3: Do 10.1.3.1 and 10.1.3.2 in any order, then EXIT)

10.1.3.1. Document historical aspects

10.1.3.2. Document current aspects

10.1.4. Identify sociological aspects (Plan 10.1.4: Do 10.1.4.1 and 10.1.4.2 in any order, then EXIT)

10.1.4.1. Document historical aspects

10.1.4.2. Document current aspects

10.2. Document negative elements of the neighborhood (Plan 10.2: Do 10.2.1 and 10.2.2 in any order, then EXIT)

10.2.1. Identify vandalism and neglect

10.2.2. Document criminal activity reports and data

10.3. Document neighborhood attitudes (Plan 10.3: Do 10.3.1, then 10.3.2, 10.3.3, and 10.3.4, respectively, then EXIT)

10.3.1. Identify attitudes about the site

10.3.2. Identify attitudes about the proposed project

10.3.3. Identify positive attitudes of the neighborhood

10.3.4. Identify negative attitudes of the neighborhood

(Continued)

Table 3.9 (*Continued*)   Full HTA Plan for Task Conduct Site Analysis

11. Assemble climatic data (Plan 11: Do 11.1, then 11.2–11.4 in any order, then EXIT)
    11.1. Document climate variation throughout the year (Plan 11.1: Do 11.1.1–11.1.4 in any order, then EXIT)
        11.1.1. Document annual temperatures (Plan 11.1.1: Do 11.1.1.1–11.1.1.4 in any order, then EXIT)
            11.1.1.1. Identify average highest temperature
            11.1.1.2. Identify average lowest temperature
            11.1.1.3. Identify average day time temperature
            11.1.1.4. Identify average night time temperature
        11.1.2. Document annual humidity levels (Plan 11.1.2: Do 11.1.2.1 then 11.1.2.2 and 11.1.2.3 in any order, then EXIT)
            11.1.2.1. Identify monthly average humidity
            11.1.2.2. Identify period of high humidity
            11.1.2.3. Identify periods of low humidity
        11.1.3. Document annual rainfall levels (Plan 11.1.3: Do 11.1.3.1 then 11.1.3.2 and 11.1.3.3 in any order, then EXIT)
            11.1.3.1. Identify average monthly rainfall
            11.1.3.2. Identify maximum rainfall
            11.1.3.3. Identify minimum rainfall
        11.1.4. Document annual snowfall levels (Plan 11.1.4: Do 11.1.4.1 then 11.1.4.2, then EXIT)
            11.1.4.1. Identify average monthly snowfall
            11.1.4.2. Identify maximum snowfall
    11.2. Document prevailing winds throughout the year (Plan 11.2: Do 11.2.1, then 11.2.2, 11.2.3, and 11.2.4, respectively, then EXIT)
        11.2.1. Identify prevailing direction each month
        11.2.2. Identify average velocity each month
        11.2.3. Identify maximum velocity expected
        11.2.4. Identify any daily variations
    11.3. Document sun path throughout the year (Plan 11.3: Do 11.3.1, then 11.3.2, 11.3.3, and 11.3.4, respectively, then EXIT)
        11.3.1. Identify summer solstice
        11.3.2. Identify winter solstice
        11.3.3. Identify altitude and azimuth days in summer (Plan 11.3.3: Do 11.3.3.1–11.3.3.4 in any order)
            11.3.3.1. Identify at sunrise
            11.3.3.2. Identify at sunset
            11.3.3.3. Identify at 0900
            11.3.3.4. Identify at 1500
        11.3.4. Identify altitude and azimuth days in winter (Plan 11.3.4: Do 11.3.4.1–11.3.4.4 in any order)
            11.3.4.1. Identify at sunrise
            11.3.4.2. Identify at sunset

(Continued)

Table 3.9 (*Continued*)   Full HTA Plan for Task Conduct Site Analysis

---

        11.3.4.3. Identify at 0900

        11.3.4.4. Identify at 1500

  11.4. Identify potential and historic natural disasters (Plan 11.4: Do 11.4.1–11.4.5 in any order, then EXIT)

      11.4.1. Document previous earthquakes and existing fault lines

      11.4.2. Document previous cyclone/hurricanes/tornadoes

      11.4.3. Document previous floods

      11.4.4. Document previous potential bushfire/wildfires risks

      11.4.5. Identify projected sea-level rises

Source: Adapted from White, E., *Site Analysis: Diagramming Information for Architectural Design*, Architectural Media, Florida, 1983.

Table 3.10   Full HTA Plan for Task-Assess Development Application

---

0. Assess development application
Plan 0: Do 1 then 2 if application does not require further assessment do 6 then EXIT, if application requires referral, information, or public notification, then do 3, 4, 5, as required, then do 6, then EXIT

1. Development application (DA) is received at development authority
2. Assessment manager (AM) reviews DA (Plan 2: Do 2.1, then 2.2, then either 2.3 or 2.4, if rejected, then EXIT, if continued, then do 2.5, 2.6, and 2.7, respectively, then EXIT)
  2.1. AM checks DA summary for properly made submission
  2.2. AM determines if DA is properly made
  2.3. AM provides confirmation of rejection of DA
  2.4. AM provides continuance of DA
  2.5. AM checks DA summary for level of assessment
  2.6. AM determines level of assessment
  2.7. AM determines if DA is referred
3. Application is referred (Plan 3: Do 3.1, if action required, do 3.2 then 3.3 and 3.4, respectively, then EXIT, if no action, do 3.4, then EXIT)
  3.1. Referral Agency (RA) checks application
  3.2. RA gives action notices
  3.3. Applicant takes required actions
  3.4. Referral agency gives confirmation

(Continued)

Table 3.10 (*Continued*)   Full HTA Plan for Task Assess Development Application

---

4. Information is requested (PLAN 4: Do either 4.1 or 4.2, if am request then do 4.3 then 4.4, then EXIT, if RA requests, do 4.3, then 4.4, and 4.5, respectively, then EXIT)
   4.1. AM requests further DA details
   4.2. RA requests further DA details
   4.3. Applicant responds to information request
   4.4. Wait for further DA details
   4.5. Referral agency issues response to AM
5. Public notification (PLAN 5: Do 5.1, then 5.2 and 5.3 in any order, then 5.4, then EXIT)
   5.1. Applicant carries out PN
   5.2. Submissions to application are received
   5.3. Applicant provides notice of compliance
   5.4. AM considers submissions
6. Decision on application (PLAN 6: Do 6.1 then 6.2 and 6.3, respectively, then EXIT)
   6.1. Review requested DA details
   6.2. Decide application
   6.3. Issue decision

4

COGNITIVE TASK
ANALYSIS METHODS

4.1 Introduction

In contrast to traditional task-analysis techniques, which provide a physical description of the activity performed within complex systems, cognitive task analysis (CTA) techniques are used to determine and describe the cognitive processes used by individuals and teams undertaking tasks. Participants undertaking activity in today's complex systems face increasing demands upon their cognitive skills and resources. As system complexity increases, they may require training in specific cognitive skills and processes to keep up. System designers, including urban designers, require an analysis of the cognitive skills and demands associated with the operation and use of complex systems. This analysis allows them to propose design concepts, allocate tasks, develop training procedures and work processes, ensure safety, and to evaluate performance and efficiency. For land use planning and urban design (LUP & UD), explicitly understanding the cognitive processes associated with the activities in urban environments is of increasing importance. Our presently aging population and the general increased incidents of cognitive impairment mean that we need to rethink many of our *business as usual* urban environments—the methods within this chapter may offer us new insights.

The past three decades have seen the emergence of CTA, and a number of techniques now exist that can be used to determine, describe, and analyze the cognitive processes employed during task performance. According to Schraagen et al. (2000), CTA represents an extension of traditional task-analysis techniques used to describe the knowledge, thought processes, and goal structures underlying observable task performance. Militello and Hutton (2000) described CTA techniques as those that focus upon describing and representing the cognitive elements that underlie goal generation, decision-making, and judgments.

CTA techniques are useful in evaluating individual and team performance within complex systems. They offer an analysis of cognitive processes surrounding decisions made and choices taken. These insights allow practitioners to design fit for purpose environments and develop guidelines for effective performance in complex urban settings. The main problem associated with the use of CTA techniques is the considerable amount of resources required. CTA methods are commonly based upon interview and observational data and therefore require considerable time and effort to conduct. Access to subject matter experts (SMEs) is also required, as is great skill on the analyst's behalf. CTA techniques are also criticized for their reliance upon the recall of events or incidents from the past. Klein and Armstrong (2004) suggested that methods that analyze retrospective incidents are associated with concerns of data reliability due to memory degradation.

4.2 The Critical Decision Method

4.2.1 Background and Applications

The critical decision method (CDM) (Klein and Armstrong, 2004) is a semi structured interview technique that uses cognitive probes to identify the cognitive processes underlying decision-making in complex environments. Typically, scenarios are decomposed into critical decision points, and so-called *cognitive probes* (targeted interview probes focusing on cognition and decision-making) are used to identify and investigate the cognitive processes underlying users and operator performance at decision points. The technique is an extension of the critical incident technique (CIT) (Flanagan, 1954) and was developed to study the naturalistic decision-making strategies of experienced personnel. The CDM procedure is perhaps the most commonly used CTA technique and has been applied in a number of domains, including the military and paramedics (Klein et al., 1989) white-water rafting (O'Hare et al., 2000), emergency response coordination (Salmon et al., 2011), and air traffic control (Walker et al., 2010).

4.2.2 Domain of Application

Generic.

4.2.3 Application in Land Use Planning and Urban Design

The CDM approach can be applied for a range of purposes, including the identification of any training or guideline requirements for the assessment, use or design of urban environments, and also for the evaluation of task performance. It is also useful for comparing the decision-making strategies employed by different users (e.g., elderly, young) of urban and regional settings or indeed (e.g., novice or expert) user analysis of land use planning policies and procedures.

4.2.4 Procedure and Advice

Step 1: Define the Task or Scenario Under Analysis
  It is first important to clearly define the aims of the analysis and to define the incident that is to be analyzed. If the scenario under analysis is not already specified, the analyst(s) may identify an appropriate incident via interview with an appropriate user or expert by asking them to describe a recent highly challenging (i.e., high workload), nonroutine incident or urban experience in which they were involved. The interviewee involved in the CDM analysis should be the primary decision maker in the chosen incident.

Step 2: Select/Develop Appropriate CDM Interview Probes
  The CDM technique works by asking participants predefined *cognitive probes* that are designed specifically to elicit information regarding the cognitive processes undertaken during task performance. It is therefore highly important that an appropriate set of probes is selected or developed prior to the analysis. The probes used are dependent upon the aims of the analysis and the domain in which the incident is embedded. Alternatively, if there are no adequate probes available, the analyst(s) can develop novel probes based upon the urban or planning analysis needs. A set of CDM probes defined by O'Hare et al. (2000) is presented in Table 4.1.

Table 4.1   CDM Probes

Goal specification	What were you aiming to accomplish through this activity?
Assessment	Suppose you were to describe the situation at this point to someone else. How would you summarize the situation?
Cue identification	What features were you looking for when you formulated your decision?
	How did you know that you needed to make the decision? How did you know when to make the decision?
Expectancy	Were you expecting to make this sort of decision during the course of the event?
	Describe how this affected your decision-making process.
Options	What courses of action were available to you? Were there any other alternatives available to you other than the decision you made?
	How/why was the chosen option selected? Why were the other options rejected?
	Was there a rule that you were following at this point?
Influencing factors	What factors influenced your decision-making at this point?
	What was the most influential factor/piece of information that influenced your decision-making at this point?
Situation awareness	What information did you have available to you at the time of the decision?
Situation assessment	Did you use all of the information available to you when formulating the decision?
	Was there any additional information that you might have used to assist in the formulation of the decision?
Experience	What specific training or experience was necessary or helpful in making this decision?
	Do you think further training is required to support decision-making for this task?
Mental models	Did you imagine the possible consequences of this action?
	Did you create some sort of picture in your head? Did you imagine the events and how they would unfold?
Decision-making	How much time pressure was involved in making the decision? How long did it actually take to make this decision?
Conceptual	Are there any situations in which your decision would have turned out differently?
Guidance	Did you seek any guidance at this point in the task/incident? Was guidance available?
Basis of choice	Do you think that you could develop a rule, based on your experience, which could assist another person to make the same decision successfully?
Interventions	What interventions do you think would prevent inappropriate decisions being made during similar incidents in the future?

Source: O'Hare, D. et al., Cognitive task analysis for decision centred design and training, in J. Annett and N. A. Stanton (Eds.), *Task Analysis*, pp. 170–190, Taylor & Francis Group, London, UK, 2000; Crandall, B. et al., *Working Minds: A Practitioner's Guide to Cognitive Task Analysis*, MIT Press, Cambridge, MA, 2006.

Step 3: Select Appropriate Participant(s)

Once the scenario under analysis and the probes to be used are defined, an appropriate participant or set of participants should be identified. They are typically the primary decision maker in the task or scenario under analysis.

Step 4: Gather and Record Account of the Incident

The CDM procedure can be applied to a scenario observed by the analyst or to a retrospective scenario described by the participant. If the CDM analysis is based upon an observed incident, then this step involves first observing the incident and then recording an account of the incident. Alternatively, the incident can be described retrospectively from memory by the participant. The analyst should ask the participant for a description of the incident in question, from its starting point to its end point.

Step 5: Construct Incident Timeline

The next step in the CDM analysis is to construct a timeline of the incident described in Step 4. The aim of this is to give the analyst(s) a clear picture of the incident and its associated events, including when each event occurred and what the duration of each event was. According to Klein et al. (1989), the events included in the timeline should encompass any physical events, or environments such as time of day or noises, and also *mental* events, such as the thoughts and perceptions of the interviewee during the incident.

Step 6: Define Scenario Phases

Once the analyst has a clear understanding of the incident under analysis, the incident should be divided into key phases or decision points. It is recommended that this is done in conjunction with the SME. Normally, the incident is divided into four or five key phases.

Step 7: Use CDM Probes to Query Participant Decision-Making

For each incident phase, the analyst should probe the participant by using the CDM probes selected during Step 2 of the procedure. The probes are used in an unstructured interview format to gather pertinent information regarding the SME's decision-making during each incident phase. The interview should be recorded using an audio-recording device.

Step 8: Transcribe Interview Data

Once the interview is complete, the data should be transcribed accordingly. It is normally useful for data-representation purposes to produce CDM tables, containing the cognitive probes and interviewee responses, for each participant.

Step 9: Construct CDM Tables

Finally, a CDM output table for each scenario phase should be constructed. This involves simply presenting the CDM probes and the associated SME answers in an output table. The CDM output tables for the use of a public open space are presented in Table 4.2.

Table 4.2   Respondent 1

Goal specification	What were you aiming to accomplish through this activity? • *Play with children, exercise the dog at a local park*
Assessment	Describe here the situation in which you made the decision about a public open space to visit • *Wanted to get children off their devices and run around and get them outside. The space needed to be big enough to run/throw rugby ball and exercise the dog at the same time*
Cue identification	What features were you looking for when you formulate your decision? • *Close proximity (walking distance) to home, grassy area, clean, safe for family, dog friendly* When do you need to make that decision? • *Knowledge of the local area* How far in advance to leaving would you make that decision? • *5 min*
Expectancy	How often would you expect to make this sort of decision? • *Three times a week* Considering that level of expectation, describe how this may affect your decision-making process • *Nil*
Options	What courses of action are available to you in making that decision? • *Local knowledge* Are there any other alternatives available to you other than the decision you made? • *Look on google/online for other local parks* How/why was the chosen option selected? Why were the other options rejected? • *Only space available to do everything we needed to do within walking distance to home*

(Continued)

Table 4.2 (*Continued*)   Respondent 1

Influencing factors	What factors influenced your decision-making? • *Time of day/if the park had lots of people there already* What was the most influential factor/piece of information that influenced your decision-making at this point? • *Space and proximity*
Situation awareness	What information did you have available to you at the time of the decision? • *Location of the park* • *Open space to play* • *Weather*
Situation assessment	Did you use all of the information available to you when formulating the decision? • *Yes* Was there any additional information that you might have used to assist in the formulation of the decision? • *Something to tell me how many people/families were already there (it was really busy)* • *I thought it was dog friendly but when I left I saw a sign that said "no dogs" on the exit (not where I had entered)*
Experience	Was there an experience that was necessary or helpful in making this decision? • *Had been there before*
Mental models	Did you imagine the possible consequences of making this choice? • *Having fun and family time* Did you create some sort of picture in your head? • *Happy and tired kids!* Did you imagine the events and how they would unfold? • *Yes*
Decision-making	How much time-pressure was involved in making the decision? • *Some—we needed to get out before dark* How long did it actually take to make this decision? • *5 min*
Conceptual	Are there any situations in which your decision would have turned out differently? • *If the weather changed* • *If the park grounds seemed unsafe*
Guidance	Did you seek any guidance at this point of the decision-making? • *Yes, the family was involved in the decision* Was guidance available? • *Yes which park to go to and what we were going to do when we got there*
Basis of choice	Do you think that you could develop a rule, based on your experience, which could assist another person to make the same decision successfully? • *Yes*

4.2.5 Advantages

- The CDM analysis procedure can be used to elicit specific information regarding the decision-making strategies used by agents in complex, dynamic systems.
- The technique is normally quick to apply.
- Once familiar with the technique, CDM is relatively easy to apply.
- It is a popular procedure and has been applied in a number of domains.
- Its output can be used to construct propositional networks that describe the knowledge or situation awareness (SA) objects required during the scenario under analysis.

4.2.6 Disadvantages

- The reliability of such a technique is questionable. Klein and Armstrong (2004) suggested that methods that analyze retrospective incidents are associated with concerns of data reliability, due to evidence of memory degradation.
- The data obtained are highly dependent upon the skill of the analyst conducting the CDM interview and also the quality of the participants used.
- A high level of expertise and training is required to use the CDM to its maximum effect (Klein and Armstrong, 2004).
- It relies upon interviewee verbal reports to reconstruct incidents. How far a verbal report accurately represents the cognitive processes of the decision maker is questionable. Facts could easily be misrepresented by the participants involved.
- It is often difficult to gain sufficient access to appropriate experts to conduct a CDM analysis.

4.2.7 Flowchart

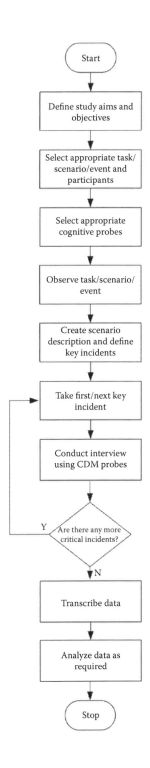

4.2.8 Related Methods

The CDM is an extension of the CIT (Flanagan, 1954). It is also closely related to other interview-based CTA techniques, in that it uses probes to elicit data regarding task performance from participants. Other similar CTA techniques include applied cognitive task analysis (ACTA) (Militello and Hutton, 2000) and walkthrough analysis (Chapter 9) (Polson et al., 1992). CDM is also used in conjunction with propositional networks (Chapter 7) to identify the knowledge objects required during the performance of a particular task.

4.2.9 Approximate Training and Application Times

Although the time taken for analysts to understand the CDM procedure is minimal, the training time is high due to the requirement for experience in interviews and for trainees to grasp cognitive psychology (Klein and Armstrong, 2004). In addition, once trained in the method, analysts are likely to require significant practice until they become proficient in its application. The application time for the CDM is medium. The CDM interview takes between one and two hours, and the transcription process takes approximately 1 to 2 h.

4.2.10 Reliability and Validity

Both the intra- and interanalyst reliability of the CDM approach are questionable. It is apparent that such an approach may elicit different data from similar incidents when applied by different analysts on separate participants. Klein and Armstrong (2004) suggested that there are also concerns associated with the reliability of the CDM due to evidence of memory degradation.

4.2.11 Tools Needed

When conducting a CDM analysis, pen and paper could be sufficient. However, to ensure that data collection is comprehensive, it is recommended that video- or audio-recording equipment is used. A set of relevant CDM probes, such as those presented in Table 4.1, is also required. The type of probes used is dependent upon the focus of the analysis.

4.2.12 Example

The following example represents the use of a public open space by different demographic users. The aim was to better understand the decision-making behind different individual's choice to use a particular public open space. The task was to *Recall a specific occurrence when you made a decision about leaving home to visit a public open space, such as a park.* The CDM probes in Tables 4.2 and 4.3 represent two respondents in Phase 1: the decision-making behind choosing a particular public open space to visit.

Table 4.3   Respondent 2

Goal specification	What were you aiming to accomplish through this activity?
	• *Recreation and relaxation. I wanted to walk somewhere with nice views*
Assessment	Describe here the situation where you made the decision about a public open space to visit
	• *I was at home and wanted to go out*
Cue identification	What features were you looking for when you formulated your decision?
	• *Somewhere I could have a decent walk, with nice views, preferably the beach. Not too busy*
	• *Somewhere close by to get a coffee*
	When do you need to make that decision?
	• *Before I get in the car*
	How far in advance to leaving would you make that decision?
	• *Maybe 15–20 min, I might have to think about which shoes to wear (e.g., runners if on a path, sandals I can remove if walking on the beach)*
Expectancy	How often would you expect to make this sort of decision?
	• *Once a week or so*
Options	What courses of action are available to you in making that decision?
	• *I can choose anywhere within short driving distance*
	Are there any other alternatives available to you other than the decision you made?
	• *Yes, I could go on a bush walk rather than to the beach, or to a general park*
	How/why was the chosen option selected? Why were the other options rejected?
	• *Better views, can get a coffee. Might see a whale*
Influencing factors	What factors influenced your decision-making?
	• *Having been to the location before and knowing it's a nice walk, knowing where to park, knowing it should not be too busy because out of school holiday period*
	What was the most influential factor/piece of information that influenced your decision-making at this point?
	• *Knowing it's a nice walk*

(Continued)

Table 4.3 (*Continued*)   Respondent 2

Situation awareness	What information did you have available to you at the time of the decision?
	• *My previous experience*
Situation assessment	Did you use all of the information available to you when formulating the decision?
	• *Probably not, just went where I have been before*
	Was there any additional information that you might have used to assist in the formulation of the decision?
	• *No*
Experience	Was there an experience that was necessary or helpful in making this decision?
	• *Yes, general past experience*
Mental models	Did you imagine the possible consequences of making this choice?
	• *I expected to get some exercise and be a bit more relaxed*
	Did you create some sort of picture in your head?
	• *No*
	Did you imagine the events and how they would unfold?
	• *No*
Decision-making	How much time-pressure was involved in making the decision?
	• *Little*
	How long did it actually take to make this decision?
	• *Not long, maybe a few seconds, once decided to go out*
Conceptual	Are there any situations in which your decision would have turned out differently?
	• *If it was raining, then might have gone somewhere with cover*
Guidance	Did you seek any guidance at this point of the decision-making?
	• *No*
	Was guidance available?
	• *Yes, could have googled places to go*
Basis of choice	Do you think that you could develop a rule, based on your experience, which could assist another person to make the same decision successfully?
	• *Not really*

4.2.13 Recommended Text(s)

Crandall, B., Klein, G., & Hoffman, R. (2006). *Working minds: A practitioner's guide to cognitive task analysis.* Cambridge, MA: MIT Press.

Klein, G., & Armstrong, A. A. (2004). Critical decision method. In N. A. Stanton et al. (Eds.), *Handbook of human factors and ergonomics methods*, (pp. 35.1–35.8). Boca Raton, FL: CRC Press.

4.3 Concept Maps

4.3.1 Background and Applications

Concept maps (Crandall et al., 2006) are used to elicit and represent knowledge via the use of networks depicting concepts and the relationships between them. Representing knowledge in the form of a network is a popular approach that has been used by cognitive psychologists for many years. According to Crandall et al. (2006), concept maps were first developed by Novak (1977; cited in Crandall et al., 2006) to understand and track changes in his students' knowledge of science. Concept maps are based on Ausubel's theory of learning (Ausubel, 1963; cited in Crandall et al., 2006) that suggests that meaningful learning occurs via the assimilation of new concepts and propositions into existing concepts and propositional frameworks in the mind of the learner. Crandall et al. (2006) pointed out that this occurs via subsumption (realizing how new concepts relate to those already known), differentiation (realizing how new concepts draw distinctions with those already known), and reconciliation (of contradictions between new concepts and those already known). Crandall et al. (2006) cited a range of studies that suggest that building good concept maps leads to longer retention of knowledge and a greater ability to apply knowledge in novel settings. Further, research using the approach has demonstrated that expertise is typically associated not only with more detailed knowledge but also with better organization of knowledge when compared with novices. The identification of knowledge requirements for different LUP & UD tasks and disciplines is an important line of enquiry for the exploration of complex urban and regional systems. An example concept map of the concept map method (Crandall et al., 2006) is presented in Figure 4.1.

4.3.2 Domain of Application

Concept maps were originally developed as an educational method for supporting meaningful learning (Ausubel and Novak, 1978; cited in Crandall et al., 2006); however, the approach is generic and can be applied in any domain. Crandall et al. (2006) cited a range of domains in which the method has been applied, including education, astrobiology, rocket science, and space exploration.

4.3.3 Application in Land Use Planning and Urban Design

In the first instance, concept maps are inherently useful in the teaching and learning of the disciplines associated with LUP & UD (e.g., town planning, urban design, engineering, architecture, landscape

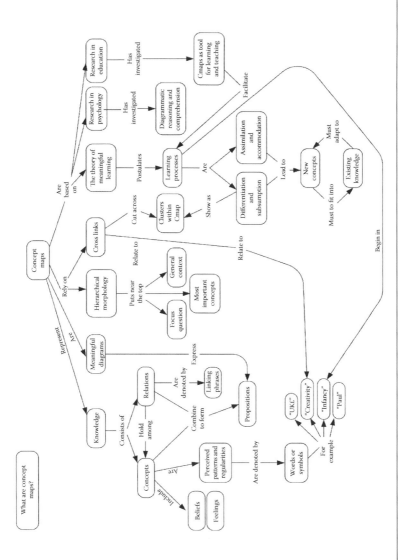

Figure 4.1 Concept map about concept maps. (Adapted from Crandall, B. et al., *Working Minds: A Practitioner's Guide to Cognitive Task Analysis*, MIT Press, Cambridge, MA, 2006.)

architecture). Concept maps may be used to elicit and represent expert knowledge; therefore, there is great scope to use them to determine and represent experienced LUP & UD practitioner's knowledge during task (design, consultation, management) performance. Concept maps can also be used to compare and contrast the knowledge of participants of differing ability.

4.3.4 Procedure and Advice

Step 1: Clearly Define Aims of the Analysis
It is first important to clearly define the aims of the analysis, so that appropriate scenarios and participants can be focused on.

Step 2: Identify Scenarios to be Analyzed
Once the aims of the analysis are clearly defined, it is next important to identify what scenarios should be analyzed. In a LUP & UD context, for example, this may be a particular design task, event, or a series of specific conceptual or development scenarios. It may also be useful to use at different design or policy development phases.

Step 3: Select Appropriate Participant(s)
Once the scenario under analysis is defined, the analyst(s) should proceed to identify an appropriate SME or set of SMEs. Typically, experts for the domain and system under analysis are used; however, if the analysis is focusing on a comparison of different users, then a selection of participants from each group should be used.

Step 4: Observe the Task or Scenario Under Analysis
It is important that the analyst(s) involved familiarizes themselves with the task or scenario under analysis. This normally involves observing the task or scenario but might also involve reviewing any relevant documentation (e.g., design guidelines, policies, standard operating procedures, existing task analyses) and holding discussions with SMEs. If an observation is not possible, a walkthrough of the task may suffice. This allows the analyst to understand the task and the participant's role during task performance.

Step 5: Introduce Participants to Concept Map Method

Crandall et al. (2006) suggested that it is important to give an introductory presentation about the concept map method to the participants involved. The presentation should include an introduction to the method, its background, an overview of the procedure, and some example applications, including a description of the methodology employed and the outputs derived.

Step 6: Identify Focus Question

Next, the knowledge elicitation and concept map construction phase can begin. Crandall et al. (2006) recommended that one analyst should act as the interviewer and one analyst act as the mapper, constructing the map online during the knowledge elicitation phase. They stress that the interviewing analyst should act as a facilitator, effectively supporting the participant in describing their knowledge during the task or scenario under analysis. This involves the use of suggestions such as "leads to?," "comes before?," and "is a precondition for?" (Crandall et al., 2006). Crandall et al. (2006) recommended that the facilitator and participant(s) should first identify a focus question that addresses the problem or concept that is to be the focus of the analysis. Examples in a LUP & UD context could be "How do you begin a site analysis?", or "Why is public art important?"

Step 7: Identify Overarching Concepts

Following the focus question, the participant should be asked to identify between five to ten of the most important concepts underlying the concept of interest (Crandall et al., 2006). These concepts should be organized in a so-called Step 1 concept map. Crandall et al. (2006) suggested that the most important or most closely related concepts should be located toward the top of the concept map.

Step 8: Link Concepts

Once the concepts are defined, the next phase involves linking them based on the relationships between them. Directional arrows and linking words are used on the concept map for this purpose. According to Crandall et al. (2006), the links between concepts can express causal relations (e.g., is caused by, results in, leads to), classificational relations (e.g., includes, refers to), property relations (e.g., owns, comprises), explanatory relations (e.g., is used for), procedure or method relations (e.g., is achieved by), contingencies and dependencies (e.g., requires), probabilistic relations (e.g., is more likely than), event relations (e.g., occurs before), and uncertainty or frequency relations (e.g., is more common than).

Step 9: Review and Refine Concept Map

The concept map is a highly iterative approach, and many revisions are normally required. The next step therefore involves reviewing and refining the map until the analysts and participant are happy with it. Refining the map might include adding concepts, subtracting concepts, adding further subordinate concepts and links, and/or changing the links. One important factor is to check that all node–link–node triples express propositions (Crandall et al., 2006).

4.3.5 Advantages

- The concept maps procedure can be used to elicit information regarding the knowledge used during task performance.
- The method is relatively easy to learn and quick to apply, at least for simple concepts and tasks.
- The approach is particularly suited to comparing the knowledge used by different users and could be applied in a LUP & UD context for this purpose.
- The method is a popular one and has been applied in a number of different domains.
- The flexibility of the approach allows all manner of concepts to be studied, including decision-making, situation awareness, error, workload, and distraction.
- The output can be used for a range of purposes, including performance evaluation, teaching materials development, and concept design analysis.
- The approach has sound underpinning theory.
- The concept map output provides a neat representation of participant knowledge.

4.3.6 Disadvantages

- The output offers little direct input into design.
- For complex concepts, the process may be difficult and time-consuming, and the output may become complex and unwieldy.
- Many revisions and iterations are normally required before the concept map is complete, even for simplistic analyses.
- The data obtained are highly dependent upon the skill of the interviewer and the quality and willingness to participate of the interviewee. A high level of skill and expertise is required to use the concept map method to its maximum effect.
- It is often difficult to gain the levels of access to SMEs that are required for the concept map method.

4.3.7 Related Methods

The concept map method is a knowledge elicitation approach that is often used in CTA efforts. Other network-based knowledge representation methods exist, including the propositional network approach (Salmon et al., 2009; Chapter 7). Concept maps use interviews and walkthrough-type analyses as their primary form of data collection. Concept maps might also be used to identify the situation awareness requirements associated with a particular task or concept.

4.3.8 Approximate Training and Application Times

The training time for the concept map method is low. The application time is also typically low, although for more complex concepts or tasks this may increase significantly. Crandall et al. (2006) suggested that typical concept map knowledge elicitation sessions take around one hour, which normally produces two semirefined concept maps.

4.3.9 Reliability and Validity

No data regarding the reliability and validity of the method are presented in the literature.

4.3.10 Tools Needed

Primarily a representational method, concept maps can be constructed simply using pen and paper, whiteboard or flipcharts; however, for the purposes of reports and presentations, it is normally useful to construct them using a drawing software package such as Microsoft Visio.

4.3.11 Example

The following concept map was developed from a verbal protocol analysis (VPA) (Chapter 3) and represents a walking task within the complex urban environment of Hong Kong. The participant has had no previous experience of this streetscape and only a basic route description to find a required public park. The aim of the study was to identify the wayfinding knowledge the participant utilized when facing a new and complex urban setting (Figure 4.2).

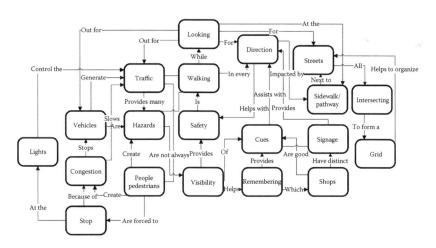

Figure 4.2 Example concept map for identifying wayfinding considerations in a new and complex urban environment.

4.3.12 Flowchart

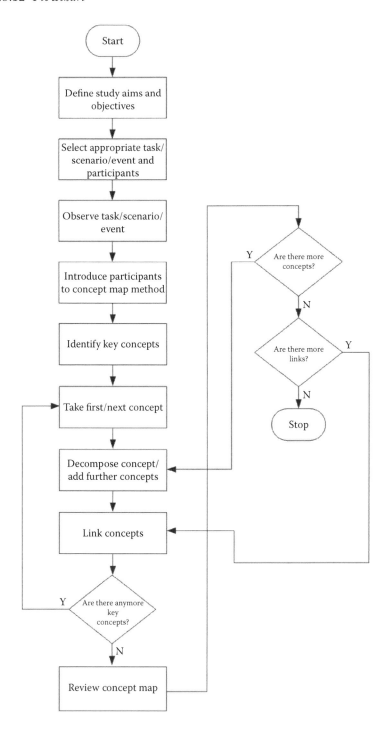

4.3.13 Recommended Text(s)

Crandall, B., Klein, G., & Hoffman, R. (2006). *Working minds: A practitioner's guide to cognitive task analysis*. Cambridge, MA: MIT Press.

4.4 Collegial Verbalization

4.4.1 Background and Applications

Collegial verbalization (Erlandsson and Jansson, 2007; Jansson et al., 2006) is a method of knowledge elicitation grounded in VPA (Ericsson and Simon, 1980). The method was designed to elicit information about participants' everyday work activities to understand their mental representations (Erlandsson and Jansson, 2007). Collegial verbalization employs a pair of experts who are closely matched in terms of experience and role. One expert is asked to perform a task while being video recorded. The second expert is then played the recording of task performance and is asked to evaluate the first expert's performance (Jansson et al., 2006). Collegial verbalization represents a hybrid method borne out of think-aloud procedures and retrospective verbalizations (Erlandsson and Jansson, 2007). McIlroy and Stanton (2011) argue that the method is both retrospective and concurrent, as verbalizations occur post-task, while the second expert (providing the verbalizations) is experiencing his or her first exposure to the task performance. The hybrid nature of the method enables it to incorporate the benefits of both methods: the minimal interference to task performance afforded by retrospective verbalizations and the reduction in rationalization afforded by concurrent verbalization methods (Erlandsson and Jansson, 2007; Jansson et al., 2006; Stanton and McIlroy, 2011).

The method provides an in-depth appreciation of task performance, including resources utilized and decision processes engaged in (Erlandsson and Jansson, 2007), as well as taking into account the impact of contextual, environmental factors on work performance (Jansson et al., 2007).

4.4.2 Domain of Application

Collegial verbalization is a generic method that can be used in any context. Previous applications of the method have explored the high-speed ferry domain (Erlandsson and Jansson, 2007) and the rail domain (Jansson et al., 2006).

4.4.3 Application in Land Use Planning and Urban Design

For LUP & UD, it has potential in assessing cognitive processes associated with site analysis and mapping. Further, it may be used to reveal the quality and value of practitioner delivery of community consultation, public participation, stakeholder engagement, and teaching.

4.4.4 Procedure and Advice (Erlandsson and Jansson, 2007)

Step 1: Define the Task Under Analysis

The first stage of analysis involves clearly defining the scenario and task under analysis. It is recommended that observations of the task are undertaken before the analysis proceeds.

Step 2: Identify and Recruit Participants

Once the task scenario has been clearly defined, the analyst must identify and then recruit expert participants who are closely matched in terms of skill, experience, and role. Access to the experts' work domain must also be arranged.

Step 3: Training in Verbalization

Next, the participants should be briefed regarding what is required of them during analysis. What the second expert should report verbally is clarified here. A small demonstration and, if required, a short practice run should be given at this stage.

Step 4: Understand the System Under Analysis

At this stage, the analyst should ensure that he or she has an understanding of the system and task under analysis. Jansson et al. (2006) posited that the analyst should observe the task under analysis before the trial to identify appropriate positions for video-recording equipment and so forth.

Step 5: Begin Scenario and Record Data

At this stage, the participant should be asked to conduct the task (or tasks) under analysis. The analyst should set up the video-recording equipment before the task begins and ensure that this does not interfere with the performance of the task. The participant should then be recorded undertaking the task.

Step 6: Capture Expert Verbalizations

Once the task is complete, the video recordings are shown to a second participant, a work colleague. This second participant is asked to verbalize the cognitive processes undertaken at each stage of the task performance. The analyst should ensure that this stage of the analysis is also recorded, either in a visual or auditory manner.

Step 7: Transcribe Data

All data captured during both the task performance of expert one and the verbalizations of expert two should be transcribed by the analyst.

Step 8: Interview Experts

It is suggested by Erlandsson and Jansson (2007) that a semi-structured interview may be beneficial at this stage to discuss the findings with the experts.

Step 9: Analyze Data

The final stage of the procedure involves collating the transcripts and drawing out key insights concerning the cognitive processes captured.

4.4.5 *Advantages*

- The method does not interfere with task performance, unlike think-aloud procedures, which removes the additional workload associated with verbalizations and prevents any disruption to task performance (Erlandsson and Jansson, 2007).
- The use of a second expert prevents participants from *rationalizing* their actions (Erlandsson and Jansson, 2007).
- The method allows for the description of processes that may have become automated (Erlandsson and Jansson, 2007).
- It provides data with a higher level of detail and reliability than think-aloud or retrospective methods (Erlandsson and Jansson, 2007).
- It enables the elicitation of implicit knowledge and exploration of mental models of the system (Erlandsson and Jansson, 2007).
- The use of two experts provides a level of validity to the results over those based on a single participant (Erlandsson and Jansson, 2007).

- It removes the need for subjective interpretation by the analyst (Jansson et al., 2006).
- It provides useful information for system design (Jansson et al., 2006).

4.4.6 Disadvantages

- The use of video-recording equipment could influence task performance (McIlroy and Stanton, 2011); this may be especially true when participants know that colleagues will later evaluate their performance.
- McIlroy and Stanton (2011) argued that the method separates the verbalizations from the task performance through the utilization of two participants.
- As verbal protocols are not provided by the person conducting the tasks, there is the possibility that the protocols may not exactly match the cognition of the person (Erlandsson and Jansson, 2007).
- Erlandsson and Jansson (2007) argued that further investigation is needed to validate the methodology.
- Due to the naturalistic approach taken by this method, events that are not common may not be captured (Erlandsson and Jansson, 2007).
- It provides a large amount of data to be analyzed.

4.4.7 Related Methods

Collegial verbalization is similar to elicitation by critiquing (EBC: Miller et al., 2006). EBC involves asking participants to think aloud while performing a task, and both task performance and the concurrent verbalizations are recorded. Post-task performance, an expert is presented with the recordings and is asked to evaluate the participant's performance. This expert evaluation is also recorded and studied by the analyst. The only difference between EBC and collegial verbalization is that EBC involves both experts providing verbalizations, rather than only the second expert providing verbalizations.

The concurrent observer narrative technique (CONT) (McIlroy and Stanton, 2011) is another method similar in approach to collegial verbalization, which utilizes work colleagues to provide verbalizations of task performance. Within CONT, the work colleague providing the verbalizations is observing task performance from within the work environment as opposed to the video recordings of task performance utilized in collegial verbalization.

Video-cued recall procedure (Omodei and McLennan, 1994, cited in Erlandsson and Jansson, 2007) is an additional knowledge elicitation technique, which involves a video recording of the participant's task performance, recorded using a head camera. Post-task performance, participants watch the recordings and are asked to provide verbalizations to describe their actions at each stage of performance.

4.4.8 Approximate Training and Application Times

Previous applications of the method have not listed application times; however, it is anticipated that the large amount of data collected by the method would lead to considerable data analysis. The use of two experts will significantly increase the application time of the method.

As with all VPAs, time will need to be spent for training participants in appropriate verbalization techniques.

4.4.9 Reliability and Validity

There are no explicit data regarding reliability and validity of the method, although the authors of the method argue that increased validity is afforded by the use of multiple experts over a single participant (Erlandsson and Jansson, 2007). The method also removes reliability constraints associated with posttrial verbalizations, such as not remembering aspects of the trial accurately.

4.4.10 Tools Needed

To employ collegial verbalization, two experts, closely matched on experience and role, are required. In addition, access to the experts' work domain is required, including the ability to employ video- and audio-recording equipment.

4.4.11 Flowchart

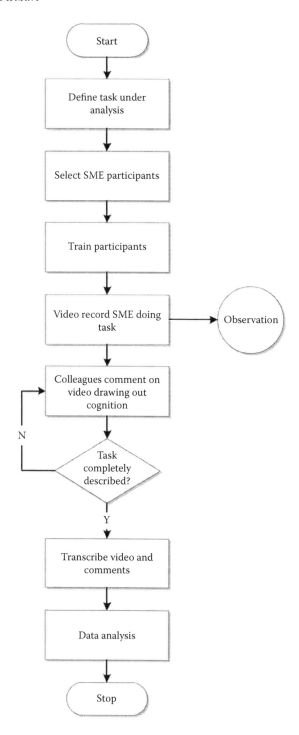

4.4.12 Recommended Text(s)

Erlandsson, M., & Jansson, A. (2007). Collegial verbalisation—a case study on a new method on information acquisition. *Behaviour & Information Technology, 26*(6), 535–543.

5

ERROR IDENTIFICATION METHODS

5.1 Introduction

Although there is now less emphasis on human error as it is seen as a consequence rather than a cause of incidents, it remains a key concept in accident analysis and prevention. Error is a complex construct that has received considerable attention from the human factors (HF) community and requires a similar level of attention from land use planning and urban design (LUP & UD). It has been consistently identified as a contributory factor in a high proportion of incidents in complex, dynamic systems at the land use and transport planning interface. For example, within the rail-transport field, (human) error was identified as a contributory cause of almost half of all collisions occurring on the UK rail network between 2002 and 2003 (Lawton and Ward, 2005). It has also been estimated that human or driver error contributes to as much as 90 percent of road accidents (Waldrop, 2015).

Although (human) error has been investigated since the dawn of the discipline, research into the construct only increased around the late 1970s and early 1980s in response to a number of high-profile catastrophes in which human error was implicated. Major incidents such as the Three Mile Island, Chernobyl and Bhopal disasters, and the Tenerife and Papa India air disasters (to name but a few) were all attributed, in part, to human error. As a result, the construct began to receive considerable attention from the HF community and also the general public.

(Human) error is formally defined as "All those occasions in which a planned sequence of mental or physical activities fails to achieve its intended outcome, and when these failures cannot be attributed to the intervention of some chance agency" (Reason, 1990: 9).

Such a definition also perhaps offers insight to the legacy of conceptual and utopian failures of LUP & UD.

Further classifications of (human) error have also been proposed, such as the slips (and lapses), mistakes, and violations taxonomy proposed by Reason (1990), which also includes a complete description of error classifications and error theory.

The prediction of (human) error is used within complex, dynamic systems to identify the nature of potential errors and the causal factors, recovery strategies, and consequences associated with them. Information derived from (human) error identification analyses is then typically used to propose remedial measures designed to eradicate the potential errors identified. EI works on the premise that an understanding of an individual's intended (work) task and the characteristics of the technology and environment being used allows us to indicate potential errors that may arise from the resulting interaction (Stanton and Baber, 1996a). As within HF, it is proposed here that EI techniques can be used during planning and urban design processes to highlight both potential design-induced error or to evaluate error potential in existing systems. These are typically conducted on a task analysis of the activity under investigation. The output of EI techniques usually describes potential errors, their consequences, recovery potential, probability and criticality, and offers associated design remedies or error-reduction strategies.

5.1.1 Error Classifications

At the most basic level of error classification, a distinction between errors of omission and errors of commission is proposed. Errors of omission are those instances in which an actor fails to act at all, whereas errors of commission are those instances in which an actor performs an action incorrectly or at the wrong time. Payne and Altman (1962; cited in Isaac et al., 2002) proposed a simplistic information-processing theory-based error classification scheme containing the following error categories:

1. *Input errors*—Those errors that occur during the input sensory and perceptual processes, for example, visual perception and auditory errors

2. *Mediation errors*—Those errors that occur or are associated with the cognitive processes employed between the perception and action stages

3. *Output errors*—Those errors that occur during the selection and execution of physical responses

The most commonly referred to error classification within the literature, however, is the slips and lapses, mistakes, and violations classification proposed by Reason (1990), an overview of which is presented in Sections 5.1.2 through 5.1.4.

5.1.2 Slips and Lapses

The most common form of error is slip-based errors. Slips are categorized as those errors in which the intention or plan was correct but the execution of the required action was incorrect. In an LUP & UD context, examples of slip-based errors would be when a practitioner undertakes a process of community consultation, yet schedules it at a time which conflicts with the availability of the intended participants; or more simply the user of space goes to sit on a park bench and misjudges the position and inadvertently falls. In both cases, the intention (i.e., to undertake community consultation or to sit down) was correct but the physical execution of the required action was not (i.e., failing to make contact with the participants or misjudging the distance). Slips are therefore categorized as actions with the appropriate intention followed by the incorrect execution and are also labeled action execution failures (Reason, 1990).

Lapse-based errors refer to more covert error forms that involve a failure of memory that may not manifest itself in actual behavior (Reason, 1990). Lapses typically involve a failure to perform an intended action or forgetting the next action required in a particular sequence. Examples of lapses within the LUP & UD context include a failure to provide or collect required information to support design decision-making or a planner failing to check the necessary site access prior to commencing a site visit. Although slips occur at the action execution stage, lapses occur at the storage stage, whereby intended actions are formulated prior to the execution stage of performing them.

5.1.3 Mistakes

Although slips reside in the observable actions made by operators, mistakes reside in the unobservable plans and intentions that they form and are categorized as an inappropriate intention or wrong decision followed by the correct execution of the required action. Mistakes occur when actors intentionally perform a wrong action and therefore originate at the planning level, rather than the execution level (Reason, 1990). There is a long list of, big and small, mistakes such as rife LUP & UD, the choice of the wrong materials, lack of individuality in design, prioritizing cars over all other transport, dispersal of employment, and developing in flood plains, not including community decision-making. According to Reason (1990), mistakes involve a mismatch between the prior intention and the intended consequences and are likely to be more subtle, more complex, less well understood, and harder to detect than slips. Reason (1990) defines mistakes as "deficiencies or failures in the judgmental and/or inferential processes involved in the selection of an objective or in the specification of the means to achieve it, irrespective of whether or not the actions directed by this decision-scheme run according to plan" (Reason, 1990).

5.1.4 Violations

Another, altogether more complex category of error is violations. Violations are categorized as any behavior that deviates from accepted procedures, standards, and rules. Violations can be either deliberate or unintentional (Reason, 1997). Deliberate violations occur when an actor deliberately deviates from a set of rules or procedures. This form of violation is also rife in LUP & UD, whereby regulation and rules are made and then bent and broken—concessions are provided and conditions eroded to provide for powerful campaigners. There are also of course erroneous or unintentional violations, whereby practitioners inadvertently go outside the boundaries of the law or break the rules.

In addition to the simplistic slips and lapses, mistakes, and violations classification described earlier, further error types have been specified within each category. For example, Reason (1990) proposed a taxonomy of unsafe acts, which prescribes a number of different error types within each of the four error categories. The taxonomy of unsafe acts is presented in Figure 5.1.

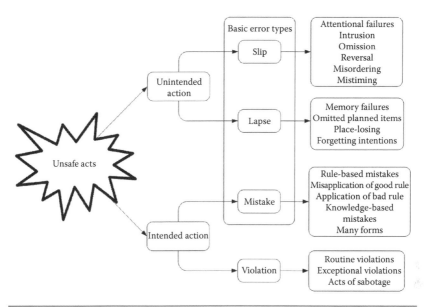

Figure 5.1   Unsafe acts taxonomy. (From Reason, J., *Human Error*, Cambridge University Press, New York, 1990.)

5.2 The Systematic Human-Error Reduction and Prediction Approach

5.2.1 Background and Applications

The most popular of all EI approaches is the systematic human-error reduction and prediction approach (SHERPA; Embrey, 1986). Originally developed for use in the nuclear-reprocessing industry, to date SHERPA has had further application in a number of domains, including the military (Stanton et al., 2009), aviation (Stanton et al., 2009), and health care (Hughes et al., 2015). SHERPA uses an external error mode (EEM) taxonomy linked to a behavioral taxonomy and is applied to a hierarchical task analysis (HTA) of the task under analysis to predict potential human- or design-induced error. In addition to being the most commonly used of the various HEI methods available, according to the literature, SHERPA is also the most successful in terms of accuracy of error predictions.

5.2.2 Domain of Application

Despite being originally developed for use in the process industries, the SHERPA behavior and error taxonomy is generic and can be applied in any domain involving human activity.

5.2.3 Application in Land Use Planning and Urban Design

In LUP & UD, it has the potential to be a very powerful tool for the error analyses of existing or proposed processes and designs—for example, issues of safety or even the affordance of undesirable behaviors within a proposed or existing urban environment.

5.2.4 Procedure and Advice

Step 1: Conduct an HTA

The first step in a SHERPA analysis involves describing the task or scenario under investigation. For this purpose, an HTA of the task or scenario under analysis is normally conducted. The SHERPA technique works by indicating which of the errors from the SHERPA error taxonomy are credible at each bottom-level task step in an HTA of the task under analysis. A number of data collection techniques may be used to gather the information required for the HTA, such as interviews with subject matter experts (SMEs) and observations of the task under analysis.

Step 2: Task Classification

Next, the analyst should take the first (or next) bottom-level task step in the HTA and classify it according to the SHERPA behavior taxonomy, which is presented in the following (Stanton, 2005a):

- Action (e.g., pressing a button, pulling a switch, opening a door)
- Retrieval (e.g., getting information from a screen or manual)
- Checking (e.g., conducting a procedural check)
- Selection (e.g., choosing one alternative over another)
- Information communication (e.g., talking to another party)

For example, in the development assessment EI, the task "check development assessment summary for properly made application" would be classified as a "Check" behavior, whereas the task steps "requests further details" and "issues decision" are both "Information communication" behaviors.

Step 3: Identify Likely Errors

The analyst then uses the associated error mode taxonomy and domain expertize to determine any credible error modes for the task in question. For each credible error (i.e., those judged by the analyst to be possible), the analyst should give a description of the form that the error would take, such as "application is not properly made." The SHERPA error mode taxonomy is presented in Figure 5.2.

Step 4: Consequence Analysis

The next step involves determining and describing the consequences associated with the errors identified in Step 3.

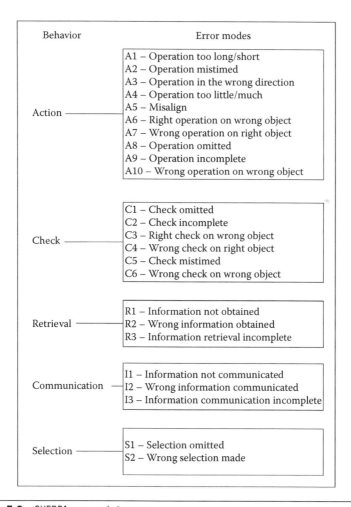

Behavior	Error modes
Action	A1 – Operation too long/short A2 – Operation mistimed A3 – Operation in the wrong direction A4 – Operation too little/much A5 – Misalign A6 – Right operation on wrong object A7 – Wrong operation on right object A8 – Operation omitted A9 – Operation incomplete A10 – Wrong operation on wrong object
Check	C1 – Check omitted C2 – Check incomplete C3 – Right check on wrong object C4 – Wrong check on right object C5 – Check mistimed C6 – Wrong check on wrong object
Retrieval	R1 – Information not obtained R2 – Wrong information obtained R3 – Information retrieval incomplete
Communication	I1 – Information not communicated I2 – Wrong information communicated I3 – Information communication incomplete
Selection	S1 – Selection omitted S2 – Wrong selection made

Figure 5.2 SHERPA error mode taxonomy.

The analyst should consider the consequences associated with each credible error and provide clear descriptions of the consequences in relation to the task under analysis. So for the error the "application is not properly made," it would have the consequences of "assessment manager confirms the rejection of the development application."

Step 5: Recovery Analysis

Next, the analyst should determine the recovery potential of the identified error. If there is a later task step in the HTA at which the error could be recovered, it is entered here. If there is no recovery step, then "None" is entered; finally, if the error is recognized and recovered immediately, the analyst enters "immediate."

Step 6: Ordinal Probability Analysis

Once the consequence and recovery potential of the error have been identified, the analyst should rate the probability of the error occurring. An ordinal probability scale of low, medium, or high is typically used. If the error has not occurred previously, a low (L) probability is assigned. If the error has occurred on previous occasions, a medium (M) probability is assigned. Finally, if the error has occurred on frequent occasions, a high (H) probability is assigned.

Step 7: Criticality Analysis

Next, the analyst rates the criticality of the error in question. A scale of low, medium, and high is also used to rate error criticality. Normally, if the error would lead to a critical incident (in relation to the task in question), it is rated as a highly critical error. For the example of the "application is not properly made," it would have a medium probability, and for the success of that application, it may be considered a critical error.

Step 8: Remedy Analysis

The final stage in the process is to propose error-reduction strategies. Normally, remedial measures comprise suggested changes to the design of the process or system. According to Stanton (2005a), remedial measures are normally proposed under the following four categories:

1. Equipment (e.g., redesign or modification of existing equipment)

2. Training (e.g., changes in training provided)
3. Procedures (e.g., provision of new, or redesign of old, procedures) and
4. Organizational (e.g., changes in organizational policy or culture)

5.2.5 Advantages

- The SHERPA technique offers a structured and comprehensive approach to the prediction of human error.
- The SHERPA taxonomy prompts the analyst for potential errors.
- According to the HF literature, SHERPA is the most promising HEI technique available. It has been applied in a number of domains with considerable success. There is also a wealth of encouraging validity and reliability data available.
- It is quick to apply compared with other HEI techniques.
- It is also easy to learn and apply, requiring minimal training.
- It is exhaustive, offering error reduction strategies in addition to predicted errors, associated consequences, probability of occurrence, criticality, and potential recovery steps.
- The SHERPA error taxonomy is generic, allowing the technique to be used in a number of different domains.

5.2.6 Disadvantages

- SHERPA can be tedious and time-consuming for large, complex tasks.
- The initial HTA adds additional time to the analysis.
- It only considers errors at the *sharp end* of system operation and does not consider system or organizational errors.
- It does not model cognitive components of error mechanisms.
- Some predicted errors and remedies are unlikely or lack credibility, thus posing a false economy (Stanton, 2005a).
- Its current taxonomy lacks generalizability (Stanton, 2005a).
- It is a subjective method based on analysts' ability (Phipps et al., 2008).
- It is unable to explore contextual factors (Phipps et al., 2008).

5.2.7 Example

In the following example, SHERPA was applied to identify the errors that drivers could make at existing rail level crossings (RLX) in Victoria, Australia (Read et al. 2017). The project was concerned with improving safety at these critical interfaces within transport planning and identifying initial design ideas for improving behavior and safety.

Initially, one analyst used SHERPA to identify potential errors. This involved taking each bottom-level task step from the HTA, classifying it into one of the five SHERPA behaviors (action, check, retrieval, communication, and selection) and then using the error mode taxonomy in Figure 5.2 to identify credible errors. For each credible error, a description of the error and its consequences were documented along with any recovery steps (i.e., the point in the HTA at which the error could be recovered), ratings of probability and criticality of the error, and potential remedial measures. Following the initial analysis, two analysts with experience in applying SHERPA reviewed the analysis to check the credibility of the errors and their associated probability and criticality ratings. An extract of the RLX SHERPA is presented in Table 5.1.

A total of 92 potential errors were identified, which were fairly evenly distributed across three categories: action errors (e.g., driver fails to slow, RLX fails to activate warnings); checking errors (e.g., driver fails to look at flashing light assembly, train driver fails to look for road users); and retrieval errors (e.g., driver fails to interpret flashing lights, driver misreads signage).

The following tasks had the greatest number of potential errors associated with them:

- Detect presence of train (road vehicle driver)
- Detect RLX (road vehicle driver)

Moreover, a greater proportion of potential errors were associated with drivers, as opposed to the train driver or the technical components of the RLX (e.g., train detection device, flashing lights, boom gates). An important design implication is that new designs should exploit other components within the system such as the vehicle, technology-based detection systems, and the train itself.

Table 5.1 Extract of SHERPA for Driver Behavior at Rail Level Crossings

TASK	ERROR MODE	ERROR DESCRIPTION	CONSEQUENCE	RECOVERY	P	C	POTENTIAL REMEDIAL MEASURES
6.1.1. Look for early warning signage	C5	Driver looks for signage too late	Driver fails to comprehend approaching train	Step 6.2.1	H		In-vehicle reminder system Runway red lights in stop line (in-road studs) Alternating LED image of moving train on active early warning signage (or replace current signs with this) Traffic lights linked to RLX
6.1.2. Detect flashing lights	R1	Driver fails to detect flashing lights	Driver fails to comprehend approaching train	Step 6.3.1	M	✓	In-vehicle system Runway red lights in stop line (in-road studs) Alternating LED image of moving train on active early warning signage (or replace current signs with this) Traffic lights linked to RLX
6.1.2. Detect flashing lights	C5	Driver detects flashing lights too late	Driver fails to comprehend approaching train	Step 6.3.1	M	✓	In-vehicle system Runway red lights in stop line (in-road studs) Alternating LED image of moving train on active early warning signage (or replace current signs with this) Traffic lights linked to RLX
6.1.3. Interpret flashing lights	R1	Driver fails to interpret flashing lights (LBFTS)	Driver fails to comprehend approaching train	Step 6.3.1	M	✓	In-vehicle system Runway red lights in stop line (in-road studs) Alternating LED image of moving train on active early warning signage (or replace current signs with this) Traffic lights linked to RLX

Source: Read, G. J. M. et al., *Integrating Human Factors Methods and Systems Thinking for Transport Analysis and Design*, Taylor & Francis Group, Boca Raton, FL, 2017.

Note: *Rail level crossing* is abbreviated to *RLX*; P = Probability; C = Criticality; L = Low; M = Medium; H = High.

A final notable feature of the SHERPA analysis was that there appears to be few redundancies or opportunities for error identification, recovery, and mitigation in current RLX environments to cope with high criticality errors, such as the road user failing to detect flashing lights. This is especially the case with RLXs that do not currently have boom gates.

5.2.8 Related Methods

The initial data collection for SHERPA might involve a number of data collection techniques, including interviews, observation, and walkthroughs. An HTA of the task or scenario under analysis is typically used as the input to a SHERPA analysis. The taxonomic approach to error prediction employed by the SHERPA technique is similar to a number of other HEI approaches, such as human-error template (HET) (Marshall et al., 2003), human-error hazard and operability (HAZOP) (Kirwan and Ainsworth, 1992), and technique for the retrospective and predictive analysis of cognitive error in ATM (TRACEr) (Shorrock and Kirwan, 2002).

5.2.9 Approximate Training and Application Times

To evaluate the reliability, validity, and trainability of various techniques, Stanton and Young (1998) compared SHERPA with 11 other HF techniques. Based on the application of the technique to the operation of an in-car radio-cassette machine, the authors reported training times of around three hours (this is doubled if training in HTA is included). It took an average of 2 h and 40 min for people to evaluate the radio-cassette machine using SHERPA. In a study comparing the performance of SHERPA, human-error HAZOP, human error identification in systems tool (HEIST), and HET when used to predict design-induced pilot error, Salmon et al. (2002) reported that participants achieved acceptable performance with the SHERPA technique after only 2 h of training.

Harvey and Stanton (2013) applied SHERPA to the evaluation of in-vehicle computer interfaces and suggested that 2–4 h were required to collect data for the method and a further 8–10 h were needed to undertake the analysis.

5.2.10 Flowchart

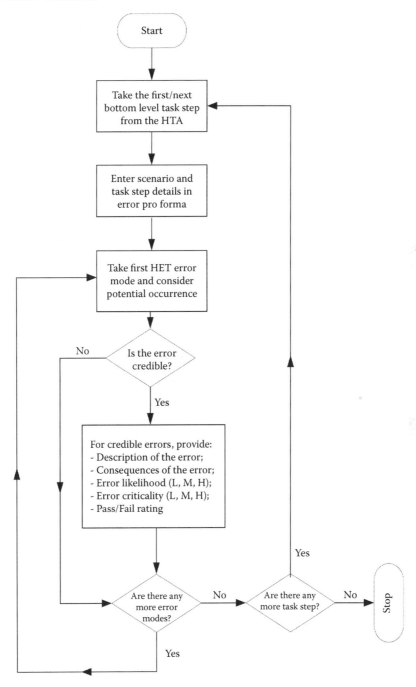

5.2.11 Reliability and Validity

There is a wealth of promising validation data associated with the SHERPA technique. Kirwan (1992) reported that SHERPA was the most highly rated of the five human-error prediction techniques by expert users. Baber and Stanton (1996a) reported a concurrent validity statistic of 0.8 and a reliability statistic of 0.9 in the application of SHERPA by two expert users to prediction of errors on a ticket-vending machine. Stanton and Stevenage (1998) reported a concurrent validity statistic of 0.74 and a reliability statistic of 0.65 in the application of SHERPA by 25 novice users to prediction of errors on a confectionery vending machine. According to Stanton and Young (1999), SHERPA achieved a concurrent validity statistic of 0.2 and a reliability statistic of 0.4 when used by eight novices to predict errors on an in-car radio-cassette machine task. According to Harris et al. (2005), SHERPA achieved acceptable performance in terms of reliability and validity when used by novice analysts to predict pilot error on a civil aviation flight scenario. Phipps et al. (2008) argued that the quality, validity, and reliability of method are dependent upon the skills of the analyst as it is a subjective method.

5.2.12 Tools Needed

SHERPA can be conducted using pen and paper. The device or at least photographs of the interface under analysis are also required.

5.2.13 Recommended Text(s)

Embrey, D. E. (1986). SHERPA: A systematic human error reduction and prediction approach. *Paper Presented at the International Meeting on Advances in Nuclear Power Systems*, Knoxville, TN.

Stanton, N. A., Young, M. S., & Harvey, C. (2014). *Guide to methodology in ergonomics: Designing for human use.* Boca Raton, FL: CRC Press.

5.3 Human-Error Template

5.3.1 Background and Applications

The HET (Harris et al., 2005) was developed for use in the certification of civil flight deck technology to predict design-induced pilot error. The impetus for HET came from a U.S. Federal Aviation

Administration report (FAA, 1996), which, amongst other things, recommended that flight-deck designs be evaluated for their susceptibility to design-induced flight crew errors and also to identify the likely consequences of those errors during the type certification process (Harris et al., 2005). The HET method is a simple checklist HEI approach that is applied to each bottom-level task step in an HTA of the task under analysis. Analysts use the HET EEMs and subjective judgment to identify credible errors for each task step. The HET EEM taxonomy consists of the following generic error modes:

- Failed to execute
- Task execution incomplete
- Task executed in the wrong direction
- Wrong task executed
- Task repeated
- Task executed on the wrong interface element
- Task executed too early
- Task executed too late
- Task executed too much
- Task executed too little
- Misread information
- Other

5.3.2 Domain of Application

The HET technique was developed specifically for the aviation domain and is intended for use in the certification of flight-deck technology. However, the HET EEM taxonomy is generic, allowing the technique to be applied in any domain.

5.3.3 Application in Land Use Planning and Urban Design

For LUP & UD, the 12-error modes of the HET error taxonomy could be applied to any task a human is expected to undertake in an urban setting. Further, it is a way in which potential errors may be identified in any number of assessment or analysis tasks that are required within LUP & UD.

5.3.4 Procedure and Advice

Step 1: Conduct an HTA

The first step in an HET analysis is to conduct an HTA of the task or scenario under investigation. The HET technique works by indicating which of the errors from the HET error taxonomy are credible at each bottom-level task step in an HTA of the task under analysis. A number of data collection techniques may be used to gather the information required for the HTA, such as interviews with SMEs and observations of the task under analysis.

Step 2: HEI

To identify potential errors, the analyst takes each bottom-level task step from the HTA and considers the credibility of each of the HET EEMs. Any EEMs that are deemed to be credible by the analyst are recorded and analyzed further. At this stage, the analyst ticks each credible EEM and provides a description of the form that the error will take.

Step 3: Consequence Analysis

Once a credible error is identified and described, the analyst should then consider and describe the consequence(s) of the error. The analyst should consider the consequences associated with each credible error and provide clear descriptions of the consequences in relation to the task under investigation.

Step 4: Ordinal Probability Analysis

Next, the analyst should provide an estimate of the probability of the error occurring, based upon his or her subjective judgment. An ordinal probability value is entered as low, medium, or high. If the analyst feels that chances of the error occurring are very small, a low (L) probability is assigned. If the analyst thinks that the error may occur and has knowledge of the error occurring on previous occasions, a medium (M) probability is assigned. Finally, if the analyst thinks that the error would occur frequently, a high (H) probability is assigned.

Step 5: Criticality Analysis

Next, the criticality of the error is rated. Error criticality is rated as low, medium, or high. If the error would lead to a serious incident (this would have to be defined clearly before the analysis), it is labeled as high. Typically, a high criticality would be associated with error consequences that would lead to substantial damage to property or people in an urban setting. If the error has consequences that still have a distinct effect on the task, such as walking in the wrong way or not being able to complete a journey, the criticality is labeled as medium. If the error would have minimal consequences that are easily recoverable, such as a small delay, the criticality is labeled as low.

Step 6: Interface Analysis

The final step in an analysis involves determining whether or not the task or activity under analysis passes the certification procedure. The analyst assigns a *pass* or *fail* rating based upon the associated error probability and criticality ratings. If a high probability and a high criticality were assigned previously, the interface in question is classed as a *fail*. Any other combination of probability and criticality and the interface in question is classed as a *pass*.

5.3.5 Advantages

- The HET methodology is quick, simple to learn and use and requires very little training.
- It utilizes a comprehensive error mode taxonomy based upon existing HEI EEM taxonomies, actual pilot error incidence data, and pilot error case studies.
- It is easily auditable as it comes in the form of an error pro forma.
- The HET taxonomy prompts the analyst for potential errors.
- The HET methodology has encouraging reliability and validity data (Marshall et al., 2003; Salmon et al., 2002; Stanton et al., 2006a).
- Although the error modes in the HET EEM taxonomy were developed specifically for the aviation domain, they are generic, ensuring that the HET technique can potentially be

used in a wide range of different domains, such as command and control, ATC, and nuclear reprocessing.

- It is a useful tool for HF certification (Stanton et al., 2006b).
- Li et al. (2009) argued that the method enables the design of a user-friendly interface that can improve task performance and increase task effectiveness.

5.3.6 Disadvantages

- For large, complex tasks, an HET analysis may become tedious and time-consuming.
- Extra work is involved if an HTA is not already available.
- It does not deal with the cognitive component of errors.
- It only considers errors at the *sharp end* of system operation and does not consider system or organizational errors.
- It is best applied by an analyst with domain-relevant knowledge (Stanton et al., 2006b).

5.3.7 Flowchart

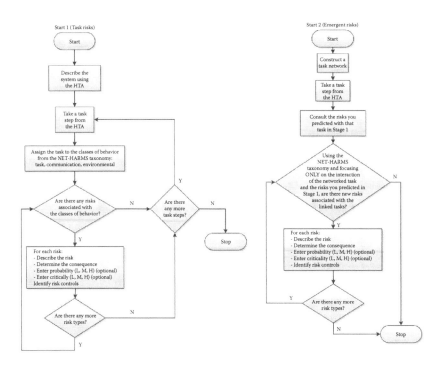

5.3.8 Example

An HET analysis, shown in Table 5.2, was conducted on the task 4.1.2. *Identify travel times to retail functions*, within the larger Site Analysis HTA (Chapter 3, page 97). The HET identified and described the possible errors, their likelihood and criticality associated with undertaking the task.

5.3.9 Related Methods

HET uses an EEM taxonomy to identify potential design-induced error. There are many taxonomic-based EI approaches available that have been developed for a variety of domains, including SHERPA, cognitive reliability and error analysis method (CREAM) (Hollnagel, 1998), and TRACEr (Shorrock and Kirwan, 2002). An HET analysis also requires an initial HTA (or some other specific task description) to be performed for the task in question. The data used in the development of the HTA may be collected through the application of a number of different techniques, including observational study, interviews, and walkthrough analysis.

5.3.10 Approximate Training and Application Times

In HET validation studies, Marshall et al. (2003) reported that with non-HF professionals, the approximate training time for the HET methodology is around 90 min. Application time varies depending on the scenario under analysis.

5.3.11 Reliability and Validity

Stanton et al. (2006a) compared HET with SHERPA, human-error HAZOP, and HEIST in their ability to predict errors during the landing of a commercial aircraft. The analysts used all methods on two separate occasions (separated by a month), and the results of these were compared with an independent collection of all errors possible during task performance (derived from interviews with, and questionnaires completed by, pilots) using a

Table 5.2   Example HET Analysis of *Identify Travel Times to Retail Functions*

SCENARIO: Urban analysis for a proposed development site. 4.3. Establish travel times between site and neighborhood functions

TASK STEP: 4.1.2. Identify travel times to retail functions

INTERFACE ELEMENT: Digital mapping

ERROR MODE	DESCRIPTION	OUTCOME	LIKELIHOOD			CRITICALITY			PASS	FAIL
			L	M	H	L	M	H		
Fail to execute	Fail to establish travel times	Proximity to development opportunities and competition will not be well understood	X					X	X	
Task execution incomplete	Active travel modes not included	No way determining understanding the walkability and cycleability			X			X		X
Task executed in the wrong direction	The travel routes do not reflect permissible access	Routes through private property may be included		X			X		X	
Wrong task executed										
Task repeated										
Task executed on the wrong interface element										
Task executed too early										
Task executed too late										
Task executed too much										
Task executed too little	Travel times are only calculated at one time of day	Peak or low times for travel time not included			X			X		X
Misread Information	Confuse the travel times for different modes	Opportunities for access to retail functions are inaccurate	X				X		X	
Other										

signal detection paradigm. The results of the study revealed that the HET analysis was statistically more accurate than the other methods, each of which was statistically comparable with one another. Stanton et al. (2006a) concluded that HET is able to predict a higher proportion of actual errors than HEIST, SHERPA, or human-error HAZOP.

5.3.12 Tools Needed

HET can be carried out using a HET error pro forma such as the headings in Table 5.2, an HTA of the task under analysis, functional diagrams of the interface under analysis, and a pen and paper.

5.3.13 Recommended Text(s)

Stanton, N., Harris, D., Salmon, P. M., Demagalski, J. M., Marshall, A., Young, M. S., Dekker, S. W. A., & Waldmann, T. (2006b). Predicting design induced pilot error using HET (Human Error Template)—A new formal human error identification method for flight decks. *Journal of Aeronautical Sciences, 110*(1104), 107–115.

5.4 The NETworked Hazard Analysis and Risk-Management System

5.4.1 Background and Applications

The NETworked hazard analysis and risk-management system (NET-HARMS) is a systems theoretic risk assessment method that combines HTA (see Chapter 3) with principles of the event analysis of systemic teamwork (see Chapter 10) and the SHERPA (see Chapter 5). The method supports the proactive identification of risk within complex sociotechnical systems by providing a description of the system under analysis on which a taxonomy is applied to identify task and emergent risks. The NET-HARMS method provides additional two key advances over existing risk assessment methods: First, it enables analysts to identify risks across the overall system, as opposed to sharp-end risks only, and, second, it enables analysts to identify emergent risks that arise when different risks from across the system interact with one another.

Applying NET-HARMS involves first developing an HTA that describes the overall system under analysis, following which task risks are identified by applying a risk mode taxonomy to each task within the HTA. Next, a task network showing the interactions between different tasks is developed, following which the risk mode taxonomy is applied once more to identify emergent risks that arise when task risks interact with one another. The output includes descriptions of the risks and their consequences, ratings of their probability and criticality, and suggested risk-management strategies.

5.4.2 *Domain of Application*

NET-HARMS was originally developed for the led outdoor activity sector (Dallat et al., In Press; Salmon et al., 2017a); however, the approach is generic in nature and can be applied in any domain for risk-assessment purposes. For example, NET-HARMS was recently used to identify task and emergent risks during railway level crossing system design and operation.

5.4.3 *Application in Land Use Planning and Urban Design*

NET-HARMS has great potential as tool to explore the risks associated with a wide variety of LUP & UD projects and processes. It can allow for the thorough analyses of the risks associated with day-to-day urban interactions, planning processes, and conceptual design. In addition, it can enable insights into the emergent risks for urban (re)development in light of changes in technology, climate, and resource-constrained urban scenarios.

5.4.4 *Procedure and Advice*

Step 1: Define Aims of the Analysis and Task or System Under Analysis
The first step in an applying NET-HARMS involves clearly defining the aims of the analysis along with the task or system under analysis. As described earlier, one of the key strengths of the NET-HARMS method is that it is capable of identifying risks across overall systems. It is therefore recommended

that NET-HARMS analyses focus on overall systems rather than frontline tasks or operations. This includes beyond the organizational level up to and including regulatory and government levels.

A useful step at this stage is to develop a clear description of the system under analysis in terms of its goals and which actors work together to achieve them. The ActorMap method (see Chapter 10) is useful for this purpose.

Step 2: Construct an HTA for the System Under Analysis

Once the task/system under analysis and aims of the analysis are clearly defined, an HTA for the system under analysis should be created. Initially, this involves collecting specific data regarding the system under analysis. Data should be collected regarding the goals and tasks involved, the human and nonhuman agents involved, the interactions between humans and nonhuman agents, the ordering of tasks and information on the factors that influence behavior. A number of different approaches can be used to collect these data, including observations, concurrent verbal protocols, structured or semistructured interviews (e.g., the critical decision method), questionnaires and surveys, walkthrough analysis, and documentation review (e.g., incident reports, standard operating procedures). The data collection approach selected is dependent upon the various constraints imposed on the analysis, such as time, number of analysts available, and access constraints.

A key focus of NET-HARMS HTAs should be on tasks that are undertaken across the system, including government and regulatory activities. Developing the HTA itself involves identifying the main goal of the system under analysis and then decomposing this into a series of sub-goals, operations, and plans. Once the initial draft HTA is complete, it is useful to have various SMEs review it. The HTA should then be refined on the basis of SME feedback. It is a normal practice for the HTA to go through many iterations before it is finalized.

Step 3: Identify Task Risks

Once the HTA is finalized, it is used first to identify task risks that might emerge during the conduct of the different tasks.

Analysts use the associated risk mode taxonomy (Table 5.3) and domain expertize to determine any credible risks for the task in question. For each credible risk (i.e., those judged by the analyst to be possible), the analyst should give a description of the form that the risk would take, such as "site selection process is driven by external pressures."

For each credible risk, analysts should determine and describe the associated consequences. For the risk of "site selection process is driven by external pressures," the consequence recorded was that "site selection is misinformed and subsequently an inappropriate development may be prioritized."

Next, analysts should provide a rating of the probability of the risk occurring. An ordinal probability scale of low, medium, or high is typically used. If the risk has not occurred previously, a low (L) probability is assigned. If the risk has occurred on previous occasions, a medium (M) probability is assigned.

Finally, if the risk has occurred on frequent occasions, a high (H) probability is assigned. The final phase of the task risk identification step involves assigning a criticality rating for each of the risks. Again, a scale of low, medium, and high is used to rate risk criticality. Normally, if the risk would lead to a critical incident (in relation to the task in question), it is rated as a highly critical risk. For the example of the "application is not properly made," it would have a medium probability,

Table 5.3   NET-HARMS Risk Mode Taxonomy

BEHAVIOR	RISK MODES
Task	T1–Task mistimed
	T2–Task omitted
	T3–Task completed inadequately
	T4–Inadequate task object
	T5–Inappropriate task
Communication	C1–Information not communicated
	C2–Wrong information communicated
	C3–Inadequate information communicated
	C4–Communication mistimed
Environmental	E1–Adverse environmental conditions

and for the success of that application, it may be considered a critical error.

Step 4: Create Task Network

Once the task risks are identified, the next step involves constructing a task network to support the identification of emergent risks. Task networks are used to represent HTA outputs in the form of a network that shows key tasks and the relationships between them (Stanton et al., 2013a). This enables analysts to understand the interactions and coupling that exists between tasks.

Within the task network, tasks are represented as nodes, and relationships are represented via lines linking the nodes. Task networks are constructed by taking each high-level task step from the HTA and identifying which tasks are related with one another. Tasks are deemed to be related with one another if the conduct of one task influences, is undertaken in combination with, or is dependent on, another task. For example, the railway level-crossing design and operation example, the tasks *select site for upgrade* and *announce upgrade program* are linked as the upgrade program announcement cannot be made until the appropriate sites are selected.

Step 5: Identify Emergent Risks

The task network is then used to support identification of emergent risks. Emergent risks represent additional risks that arise as a result of the interaction between the task risks identified during Step 3. Identifying emergent risks involves identifying the risks that arise when a task is impacted by a related task that was undertaken inadequately due to the presence of a task risk. For example, if tasks A and B are related, then analysts should determine what the impact on task B is if task A is undertaken inadequately due to the presence of a task risk. For each credible emergent risk, analysts should determine and describe the associated consequences.

Next, analysts should provide a rating of the probability of the emergent risk occurring. An ordinal probability scale of low, medium, or high is typically used. If the risk has not occurred previously, a low (L) probability is assigned. If the emergent risk has occurred on previous occasions, a medium (M) probability is assigned.

Finally, if the emergent risk has occurred on frequent occasions, a high (H) probability is assigned. The final phase of the emergent risk identification step involves assigning a criticality rating for each of the risks. Again, a scale of low, medium, and high is used to rate risk criticality. Normally, if the emergent risk would lead to a critical incident (in relation to the task in question), it is rated as a highly critical emergent risk.

Step 6: Identify Risk-Management Strategies

The final stage in the process is to propose risk-management strategies for both the task and emergent risks identified. Analysts should work through the task and emergent risk tables and discuss risk-management strategies that can be used to either prevent the risk from occurring or mitigate the consequences of the risk. Normally, risk-management strategies comprise suggested changes to the design of the process or system. According to Stanton (2005a), remedial measures proposed following risk-assessment applications are normally proposed under the following four categories:

1. Equipment (e.g., redesign or modification of existing equipment)
2. Training (e.g., changes in training provided)
3. Procedures (e.g., provision of new, or redesign of old, procedures)
4. Organizational (e.g., changes in organizational policy or culture)

Step 7: Review and Refine Analysis

Once the initial NET-HARMS is complete, it is useful to have various SMEs review it. The analysis should then be refined on the basis of SME feedback. This typically involves removing identified risks or identifying new risks or modifying probability and criticality ratings and risk-management strategies.

5.4.5 Advantages

- NET-HARMS goes beyond existing risk assessment methods to identify risks across overall systems.
- NET-HARMS identifies both task risks and emergent risks. The latter is a significant advancement over existing risk-assessment methods.
- The analysis is extremely comprehensive, covering risks across the system.
- NET-HARMS analyses provide risk-management strategies for each of the task and emergent risks identified.
- The NET-HARMS taxonomy prompts the analyst for potential task and emergent risks.
- NET-HARMS is based on SHERPA, which, according to the literature, has the most supportive reliability and validity evidence associated with it.
- It is also easy to learn and apply, requiring minimal training.
- The NET-HARMS method and taxonomy are generic, allowing it to be used in a number of different domains.
- Conducting the HTA and task network analyses enables analysts to develop an in-depth understanding of the system under analysis.

5.4.6 Disadvantages

- As it covers entire systems, NET-HARMS can be time-consuming to apply.
- The quality of the analysis is highly dependent on the expertise of the analysts.
- The requirement to develop the initial HTA and subsequent task network adds significant time to the analysis.
- Due to its infancy, there is no reliability and validity evidence associated with NET-HARMS.
- Some predicted risks and risks-management strategies may be unlikely or lack credibility, thus posing a false economy (Stanton, 2005a).

- There is currently no performance-shaping factors taxonomy to support consideration of contextual factors and their impact on task and emergent risks.

5.4.7 Related Methods

NET-HARMS requires an initial HTA of the system under analysis. The HTA is subsequently used to develop a task network for the system under analysis.

5.4.8 Approximate Training and Application Times

NET-HARMS is a relatively simple method that requires little training. In a recent reliability and validity study, participants were provided with training that lasted around one hour. Application time depends on the system under analysis; however, NET-HARMS analyses typically require significant time as the analysis covers overall systems.

5.4.9 Reliability and Validity

Although NET-HARMS has recently been subject to reliability and validity studies, the findings have not yet been published. The SHERPA method on which it is based has a significant body of promising validation evidence associated with it (Baber and Stanton, 1996a; Kirwan, 1992; Stanton and Stevenage, 1998; Stanton and Young, 1999). HTA also has some validation evidence associated with it (Stanton and Young, 1999).

5.4.10 Tools Needed

NET-HARMS can be conducted using pen and paper; however, analyses are normally undertaken using Microsoft Word. The HTA tool can be used for the HTA component, and task networks are typically drawn in Microsoft Visio.

5.4.11 Flowchart

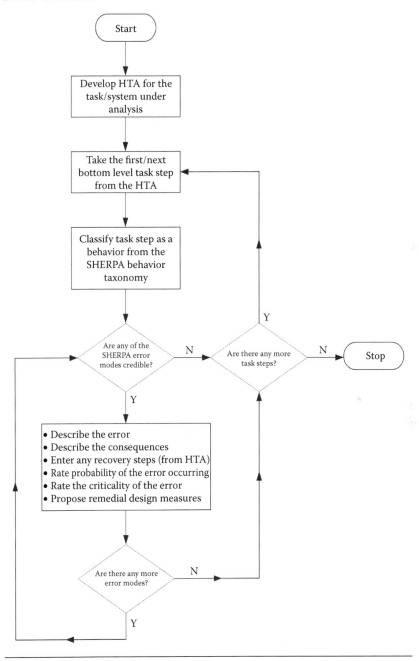

Adapted from Dallat, C. et al. *Safety Science*, 2017.

5.4.12 Example

NET-HARMS was used to identify risks arising during the design and operation of railway level crossings in Victoria, Australia. Initially, an HTA of the railway level-crossing system lifecycle was developed on the basis of various sources of information, including analyses of railway level-crossing systems (e.g., Mulvihill et al., 2016; Salmon et al., 2016a) and documentation review (e.g., road and rail safety strategy and policy documents, crash investigation reports, relevant academic literature). The resulting HTA described the goals, sub-goals, and operations required when designing, implementing, operating, and removing RLXs in Victoria, Australia. The high-level goals from the HTA are presented in the following:

1. Select sites for upgrade
2. Announce upgrade program
3. Identify user community of RLX
4. Identify RLX type and infrastructure
5. Design upgrade
6. Construction and commissioning of safe and efficient RLX
7. Manage infrastructure during construction process
8. Meet legislative requirements around risk management for RLX
9. Operate railway level crossing
10. Monitor performance
11. Check compliance with legislation
12. Grade separation
13. Crossing closure
14. Continuous improvement of standards
15. Budget allocated to level crossing program

A workshop was then held to identify the task and emergent risks. This involved four analysts working through the HTA and task networks to identify credible task and emergent risks that might impact the design and operation of railway level crossings.

For the task risk, the analysts used the NET-HARMS taxonomy to identify credible risks that could occur during completion of the railway level-crossing design and operation tasks. For each credible task risk, the analysts recorded a description of the risk, identified and recorded the consequences, and provided a rating of the probability and criticality of the risk occurring. An extract of the railway level-crossing design and operation task risks is presented in Table 5.4.

Following identification of the task risks, the HTA was used to develop a task network showing the interrelations between railway level-crossing design and operation tasks. The resulting task network is presented in Figure 5.3.

The analysts then used the NET-HARMS taxonomy and task network to identify credible emergent risks that could occur as a result of the interaction of task risks across the railway level-crossing design and operation tasks. This involved examining each set of related tasks in the task network and identifying what the impact of task risks would be on related tasks.

For each credible emergent risk, the analysts recorded a description of the risk, identified and recorded the consequences, and provided a rating of the probability and criticality of the risk occurring. An extract of the railway level-crossing design and operation emergent risks is presented in Table 5.5.

5.4.13 Recommended Text(s)

Dallat, C., Salmon, P. M., & Goode, N. (2017). Identifying risks and emergent risks across sociotechnical systems: The Networked hazard analysis and risk management system (NET-HARMS). *Theoretical Issues in Ergonomics Science*.

Dallat, C., Salmon, P. M., & Goode, N. (2017). Risky systems versus risky people: To what extent do risk assessment methods consider the systems approach to accident causation? A review of the literature. *Safety Science, (In press)*.

Stanton, N. A., & Bessell, K. (2014). How a submarine returns to periscope depth: Analysing complex socio-technical systems using cognitive work analysis. *Applied Ergonomics*, 45(1), 110–125.

Table 5.4   Extract of Railway Level–Crossing System Lifecycle Task Risks

TASK	RISK MODE	TASK RISK DESCRIPTION	TASK RISK CONSEQUENCES
15. Budget allocated to rail level–crossing upgrade program	T1	Budget allocation is delayed	Delay in upgrade of risky RLXs
	T2	Budget is not allocated	Risky RLXs not upgraded
	T4	Insufficient budget is allocated	All risky RLXs may not be upgraded
	C3	Insufficient information communicated to stakeholders regarding upgrade program	Stakeholders are unclear on upgrade program
	C4	Budget allocation communication is delayed	Delay in upgrade of risky RLXs
1. Select site for upgrade	T1	Sites selection delayed	Delay in upgrade of risky RLXs
	T3	Inappropriate sites are selected for upgrade	Risky RLXs may not be upgraded/inappropriate RLXs are upgraded
	T4	Tools to support site selection are inadequate/inappropriate	Site selection is misinformed (e.g., ALCAM/Kerang/Inappropriate RLXs are upgraded
	T5	Site selection process is driven by external pressures	Site selection is misinformed (e.g., ALCAM/Kerang/Inappropriate RLXs are upgraded
	C3	Incomplete/inadequate information is communicated regarding site selection	Upgrade may be inappropriate
	C4	Site selection communication is delayed	Delay in upgrade of risky RLXs
2. Announce upgrade program	T1/C4	Announcement is made too early	Places time pressure on the site-selection process

(*Continued*)

Table 5.4 (*Continued*) Extract of Railway Level–Crossing System Lifecycle Task Risks

TASK	RISK MODE	TASK RISK DESCRIPTION	TASK RISK CONSEQUENCES
3. Identify and consult with RLX user community	T1/C4	Announcement is delayed	Delay in upgrade of risky RLXs
	T1	Identify and consult with community too late	Upgrade and design does not take into account end-user need
	T2	Failure to identify and consult with community	Upgrade and design does not take into account end-user need
	T3	Consultation is inadequate (e.g., language and translation of information)	Upgrade and design does not take into account end-user need
	T4	Consultation methods are inadequate, for example, not suitable for user community	Upgrade and design does not take into account end-user need
	T5	Consultation methods are inappropriate, for example, sausage sizzle	Upgrade and design does not take into account all end-user need
	C1	Information not communicated or gathered	Upgrade and design does not take into account all end-user need
	C2	The wrong information is communicated or gathered	End-users needs not understood
	C3	Consultation methods are inadequate, for example, not suitable for user community	Upgrade and design does not take into account end-user need
	C4	Identify and consult with community too late	Upgrade and design does not take into account end-user need

(*Continued*)

Table 5.4 (Continued) Extract of Railway Level–Crossing System Lifecycle Task Risks

TASK	RISK MODE	TASK RISK DESCRIPTION	TASK RISK CONSEQUENCES
4. Determine infrastructure required	T1	Type of upgrade and infrastructure required are determined too late	Upgrade may not be possible/delay in upgrade of risky RLXs
	T3	Determining upgrade and type of infrastructure are done inappropriately	Wrong upgrade/infrastructure is selected
	T4	Standards and legislation are inadequate	Wrong upgrade/infrastructure is selected
	T5	Inappropriate methods/information are used to determine infrastructure required	Wrong upgrade/infrastructure is selected Standards and guidelines are violated
	C2	Wrong information communicated regarding infrastructure and upgrade required	Design documentation includes wrong information regarding upgrade
	C3	Information communicated regarding infrastructure and upgrade required is inadequate (e.g., missing detail)	Design documentation includes wrong information regarding upgrade
	C4	Information communication delayed	Design and tender process is delayed
		Information communicated too early (e.g., before legislation change)	Design decisions do not comply with current standards
5. Design upgrade and issue tender	T1	Upgrade design delayed	Delay in upgrade of risky RLXs
	T1	Tender issue delayed	Delay in upgrade of risky RLXs
	T3	Design is inadequate	Design not fit for purpose
	T3	Tender is inadequate	Contractor does not understand upgrade requirements

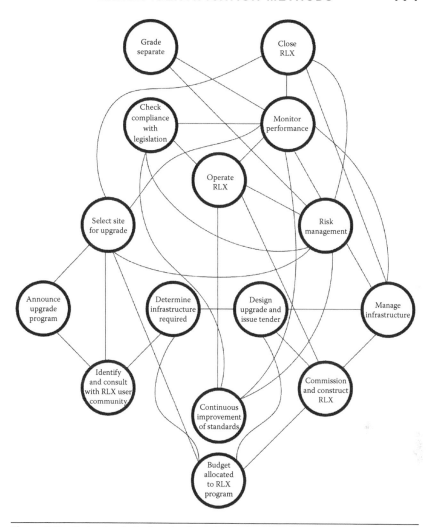

Figure 5.3   Task network of railway level-crossing system lifecycle.

Table 5.5 Extract of Railway Level–Crossing System Lifecycle Emergent Risks

TASK	RISK MODE	TASK RISK DESCRIPTION	RELATED TASK	EMERGENT RISK MODE	EMERGENT RISK DESCRIPTION	EMERGENT RISK CONSEQUENCES	P	C
15. Budget allocated to rail level crossing upgrade program	T1	Budget allocation is delayed	1. Select site for upgrade	T1	Site selection process is delayed	Inadequate railway level crossings remain in service for prolonged period of time	H	H
	T4	Insufficient budget is allocated	1. Select site for upgrade	T3	Site selection is limited to fewer sites	Inadequate railway level crossings remain in service for prolonged period of time	H	H
	T1	Budget allocation is delayed	6. Commission and construct railway level crossing	T1	Commissioning and construction of railway level crossing/design upgrade is delayed	Inadequate railway level crossing remains in service for prolonged period of time	H	H
	T2	Budget is not allocated	6. Commission and construct railway level crossing	T2	New railway level crossing/design upgrade is not commissioned	Inadequate railway level crossing remains in service for prolonged period of time	L	H
	T4	Insufficient budget is allocated	6. Commission and construct railway level crossing	T3	All required railway level crossings/design upgrades are not implemented	Inadequate railway level crossings remain in service for prolonged period of time	H	H
	C4	Budget allocation communication is delayed	6. Commission and construct railway level crossing	T1	Commissioning and construction of railway level crossing/design upgrade is delayed	Inadequate railway level crossing remains in service for prolonged period of time	H	H

6

ACCIDENT ANALYSIS
METHODS

6.1 Introduction

Although both error identification and error analysis methods can be used to analyze accidents, a subsection of methods designed specifically to focus on accident analysis is established in human factors (HF). Accident analysis methods are employed to derive an accident or incident etiology and identify contributory factors in the deviation from anticipated or safe performance. Salmon et al. (2010) identified over 30 accident analysis-related methods, illustrating the prominence of accident analysis methods in contemporary HF.

Two accident analysis methods are introduced in this chapter to provide land use planning and urban design (LUP & UD) researchers and practitioner's new ways of exploring planning and urban development as contributor factors in accidents and disaster. There are a range of circumstances in which LUP can be attributed to both the cause and indeed the exacerbation of adverse incidents, including flooding and failures of critical infrastructure. Further, such approaches may offer ways to unpick some of the complex and detrimental outcomes to human health when considering LUP's contributions to chronic disease, congestion, or indeed personal and societal safety and security. A brief introduction to these analysis methods is provided in the following.

Accimap is a method that uses a systemic control structure (based upon Rasmussen's [1997] risk framework) to guide the analyst in the identification of errors across six systemic levels. No taxonomy of error types is included, but the physical activities of the accident scenario are used as a basis to link out to contributory factors elsewhere in the system. The systems theory accident modeling and process (STAMP) method combines a taxonomy and a systemic structure of control to aid in the identification of errors, or causal factors that were present in accidents within complex systems. The method is based upon the supposition that accidents occur due to inappropriate *energy* transfers that are allowed to evolve because of ineffective safety barriers. As such, the main focus of the method is on the identification of barriers within the system.

6.2 Accimap

6.2.1 *Background and Applications*

Accimap (Rasmussen, 1997; Svedung and Rasmussen, 2002) is an accident analysis method that is used to graphically represent the systemic causal factors involved in accidents and incidents. The Accimap method differs from typical accident-analysis approaches in that, rather than identifying and apportioning blame at the sharp end, it is used to identify and represent the causal flow of events upstream from the accident and looks at the planning, management, and regulatory bodies that may have contributed to the accident (Svedung and Rasmussen, 2002). A typical Accimap uses the following six main levels: government policy and budgeting, regulatory bodies and associations, local area government planning and budgeting (including company management, technical and operational management; physical processes and actor activities; and equipment and surroundings). Failures at each of the levels are identified and linked between and across levels based on cause–effect relations. Starting from the bottom of the graph, the equipment and surroundings level provides a description of the accident scene in terms of the configuration and physical characteristics of the landscape, buildings, equipment, tools, and vehicles involved. The physical processes and actor activities level provides a description of the failures involved at the *sharp end*. The remaining levels above the physical processes level represent all of the failures by decision makers that, in the course of the decision-making involved in their normal work context, did or could have influenced the accident flow during the first two levels. Accimap analysis focuses upon the causal relationships between these levels, which allows for a vertical analysis across the levels of a system rather than the horizontal generalization within individual levels that is normally found in accident analysis methods (Svedung and Rasmussen, 2002).

6.2.2 Domain of Application

Accimap analysis is a generic approach that has been utilized in multiple domains, including aviation accidents (Royal Australian Aviation Force, 2001), police incidents (Jenkins et al., 2010), and both rail and road accidents (Hopkins, 2005; Svedung and Rasmussen, 2002).

6.2.3 Application in Land Use Planning and Urban Design

For LUP & UD, it may be used to analyze any circumstance in which, for example, LUP or urban design may have been, or suspected to be, a contributory factor within an accident or incident. Indeed, if reporting exists for an LUP & UD investigation or incident (e.g., flood report), Accimap is a very useful tool to explore the broader contributory system. Further, the ActorMap component, Step 2 (detailed in Chapter 10) has been proven as a useful way to undertake stakeholder and policy analysis of urban development and planning stakeholders and their strategic positions.

6.2.4 Procedure and Advice

Step 1: Data Collection

Being a retrospective approach, the Accimap approach is dependent upon accurate data regarding the incident under analysis. The first step therefore involves collecting data regarding the incident in question. Data collection for Accimaps can involve a range of activities, including interviews with those involved in the incident or subject matter experts (SMEs) for the domain in question, analyzing reports or inquiries into the incident, and observing recordings of the incident. As the Accimap method is so comprehensive, the data collection phase is typically time-consuming and involves analyzing numerous data sources.

Step 2: Construct an Actor Map (see also Chapter 10)

Once the data collection is complete, the analyst should identify all actors involved in the scenario and annotate these onto an actor map. Actors should be linked to one another to reflect the communication structure of the system.

Step 3: Identify Physical Process/Actor Activities Failures

After the key actors have been identified, the analyst can begin to develop the Accimap. The first stage of the Accimap involves identifying the errors involved and identifying the links between these errors. This step is concerned with the identification of errors that occurred at the level of physical process and actor activities.

Step 4: Identify Causal Factors

The analyst now needs to identify causal factors for each of the physical and actor failures identified in Step 3. Each failure is taken in turn, and the analyst identifies all related failures at the remaining five levels of the Accimap: government policy and budgeting, regulatory bodies and associations, local area government planning and budgeting, physical processes and actor activities, and equipment and surroundings.

Step 5: Identify Failures at Other Levels

Once Step 4 has been completed, the analyst should review the six systemic levels to ensure that all relevant failures have been identified. He or she should also take each level in turn and, whilst reviewing the data collected in Step 1, ensure that no failures have been missed on the Accimap.

Step 6: Finalize and Review Accimap Diagram

The Accimap should be constructed whilst the analyst steps through these stages. At this stage, the analyst should review the Accimap and ensure that all links between causal factors have been identified and that all annotated links are appropriate. SMEs should be asked to review the Accimap to ensure its validity. This review and revise stage of the Accimap process normally requires several iterations.

6.2.5 Advantages

- Accimaps enable the identification of system-wide errors that led to the occurrence of the accident at the sharp end. The complete accident etiology is exposed.
- The method is simple to learn and use.
- It is based upon a sound theoretical model.
- It considers causal factors across systemic levels.
- Its output offers an exhaustive analysis of accidents and incidents.
- It provides a clear visual interpretation of the accident etiology.
- It is a generic approach that has been applied across many domains.
- Considering the different levels involved in the accident enables an extended timeline of causality to be established.
- It focuses on systematic improvements rather than focusing on blaming individuals.

6.2.6 Disadvantages

- The method can be time-consuming.
- The quality of the analysis produced is entirely dependent upon the quality of the accident report.
- Accimap analysis does not provide a method to develop corrective measures; these are based on the judgment of the analyst.
- It does not provide a structured taxonomy for error classification.
- Accimap analysis can only be used retrospectively.
- Its graphical output can become extensive and hard to decipher when used in the analysis of complex accidents.
- Almedia and Johnson (2005) argued that it is unable to adequately explore the local rationality of those involved in the accident scenario.

6.2.7 Related Methods

To conduct Accimap analysis, data collection methods such as interviews and document reviews must be utilized first. Accimap analysis is based on the model proposed by Reason (1990) and as such is related to methods such as the incident cause analysis method and the human factors analysis and classification system (HFACS) (Wiegmann and Shappell, 2003) which are also developed from this model.

6.2.8 Approximate Training and Application Times

Accimaps is a simple method to learn and apply but can become time-consuming when applied to complex systems. Estimated timescales for the method are expected to be around one to two weeks for data collection and a further week for the initial construction of the Accimap. However, the final procedural stage of review can take additional time.

6.2.9 Reliability and Validity

There are no reliability and validity data available, but the reliability of the approach is questionable due to the low level of guidance provided and the consequent reliance on the subjective judgment of the analyst to classify causal factors.

6.2.10 Tools Needed

Pen and paper are all that are required for Accimaps analysis; however, Microsoft Office tools such as Visio can be used to create more professional Accimaps.

6.2.11 Example

The following Accimap example in Figure 6.1 represents the Murrindindi bushfire response during the devastating February 2009 bushfires in Victoria, Australia (Figure 6.1). It was developed from the *Victorian Royal Bushfires Commission, 2010. Final report, vol. II—fire preparation, response and recovery*. It represents an analysis within which LUP policies and management practices were identified as contributing factors to the disaster (Salmon et al., 2014b).

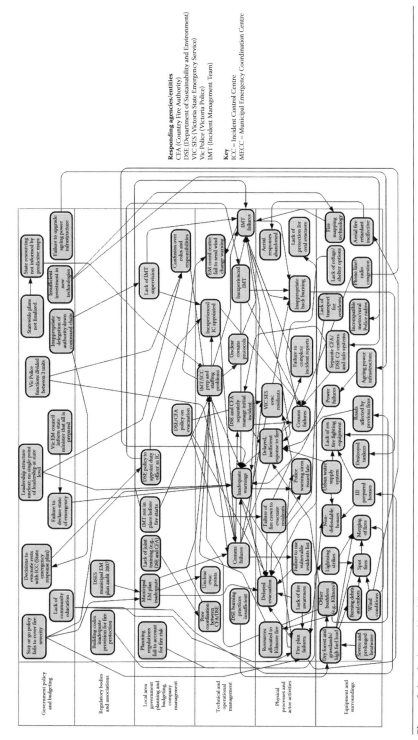

Figure 6.1 Accimap of bushfire response. (From Salmon, P. M. et al., *Safety Science*, 70, 114–122, 2014b.)

6.2.12 Flowchart

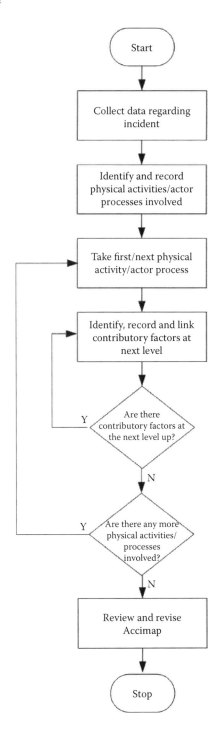

6.2.13 Recommended Text(s)

Rasmussen, J. (1997). Risk management in a dynamic society: A modelling problem. *Safety Science, 27*(2/3), 183–213.
Svedung, J., & Rasmussen, J. (2002). Graphic representation of accident scenarios: Mapping system structure and the causation of accidents. *Safety Science, 40*, 397–417.

6.3 The Systems Theory Accident Modeling and Process

6.3.1 Background and Applications

STAMP (Leveson, 2004) is an accident-analysis technique developed to explore accident causation from a systems theory perspective. Leveson argues against traditional event chain models of accident causation and instead posits that accident causation should account for processes such as adaptation and emergence that are present within complex systems. The method is based upon the hypothesis that accidents occur due to "external disturbances, component failures, or dysfunctional interactions among system components" (Leveson, 2004: 250) that are not sufficiently constrained or controlled by the system. Leveson posits that a system is made up of multiple levels that interact in unpredictable ways; to prevent an accident occurring, sufficient control should be enforced upon the system to prevent unsafe evolution.

The method provides an organizational hierarchy, control structure and taxonomy of causation, the classification of flawed control, which is applied to each level of the hierarchy to identify where the dysfunctional interactions occurred within the system under analysis (Leveson, 2004).

6.3.2 Domain of Application

The method is generic and has been applied to a number of complex systems, including aviation safety systems (Allison et al., 2017), friendly fire incidents (Leveson, 2002), the contamination of a water supply (Leveson, 2004), and road safety (Salmon et al., 2016).

6.3.3 Application in Land Use Planning and Urban Design

The STAMP approach is interested in describing various forms of control, including managerial, organizational, physical, operational, and manufacturing-based controls. Accordingly, STAMP is inherently useful to explore LUP & UD via control structure models that incorporate a series of hierarchical system levels and describe the actors and organizations that reside at each level. Control and feedback loops afford the ability to show what control mechanisms are enacted down the hierarchy and what information about the status of the system is sent back up the hierarchy.

6.3.4 Procedure and Advice

Procedure and Advice (Leveson, 2004)
  Step 1: Define the Task Under Analysis
       The initial stage of analysis involves the analyst clearly defining the task under analysis. A definition of the core goals and boundaries should be developed to guide analysis and ensure that the investigation is appropriate and relevant.
  Step 2: Data Collection
       Like all accident analysis methods, STAMP is dependent upon accurate data regarding the accident in question. The next step therefore involves collecting detailed data regarding the accident and about the domain and organization in which the accident took place. Data collection tools such as reviewing accident reports, inquiry reports and task analyses of the system in question, interviewing personnel involved in the accident, reviewing documents regarding the domain in question (e.g., rules and regulations, standard operating procedures), and/or interviewing SMEs for the domain/system in question should all be utilized.
  Step 3: Construct Hierarchical Control Structure
       By using the data collected in Step 2, the analyst must identify the key people involved in the accident scenario. These people will be spread across the different levels of the hierarchical control structure and will include those responsible for producing guidelines, developing policies, and so on. The actors should be plotted onto a graphical illustration

of the structure at the appropriate hierarchical level. In addition to plotting actors onto the control structure, the relevant constraints between levels should be annotated onto the structure as well.

STAMP provides a generic control model to guide the analyst in the construction of their specific control structure, an example of which is presented in Figure 6.2. The left-hand side of the diagram shows the control structure for system development, whereas the right hand side of the diagram shows the control structure for system operations. The arrows between the levels represent the communications between levels that are used by levels to impose constraints on the levels below them and to provide feedback to the levels above regarding how effective the constraints are (Leveson, 2004). Although each domain/system

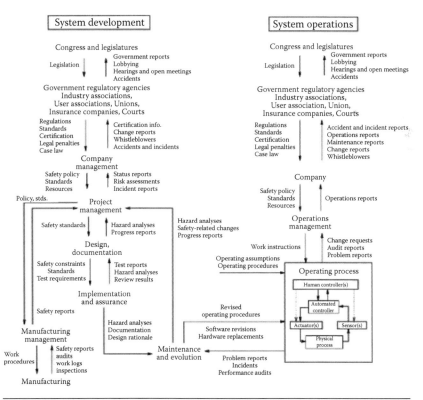

Figure 6.2 Generic control structure model. (Adapted from Leveson, N., *Safety Sci.*, 42, 237–270, 2004.)

Table 6.1   STAMP Classification of Flawed Control

1	Inadequate enforcement of constraints
1.1	Unidentified hazards
1.2	Inappropriate, ineffective, or missing control actions for identified hazards
1.2.1	Design of control algorithm (process) does not enforce constraints
	Flaws in creation process
	Process changes without appropriate change in control algorithm (asynchronous evolution)
	Incorrect modification or adaptation
1.2.2	Process models inconsistent, incomplete, or incorrect
	Flaws in creation process
	Flaws in updating process (asynchronous evolution)
	Time lags and measurement inaccuracies not accounted for
1.2.3	Inadequate coordination among controllers and decision makers
2	Inadequate execution of control action
2.1	Communication flaw
2.2	Inadequate actuator operation
2.3	Time lag
3	Inadequate or missing feedback
3.1	Not provided in system design
3.2	Communication flow
3.3	Time lag
3.4	Inadequate sensor operation (incorrect or no information provided)

Source: Leveson, N., *Safety Sci.*, 42, 237–270, 2004.

will have its own unique control structure, it is likely to be similar in structure to that presented in Figure 6.2.

Step 4: Classification of Flawed Control

Next, the analyst should take each of the systemic elements identified within the control structure and classify each according to the classification of flawed control, which is presented in Table 6.1. Analysts should apply the taxonomy to each control loop to identify the control failures involved. It is normally useful to represent the control failures identified on the control structure diagrams developed during Step 3.

Step 5: Review and Finalize Analysis

Once the first draft of the STAMP analysis is complete, the analyst should conduct a review (with SMEs if possible) to ensure that all failures have been identified and appropriately incorporated into the STAMP model. The final analysis may require several iterations.

6.3.5 Advantages

- Leveson (2004) argued that the STAMP control flaws classification scheme provides a number of different levels of analysis that are capable of exploring accident causation at a number of stages of abstraction.
- The method allows for the exploration of relationships between factors, including nonlinear relationships (Leveson, 2004).
- Leveson (2004) posited that the method can be used for accident analysis, hazard analysis, and in the development of accident prevention, safety- and risk-assessment techniques.
- The comprehensive nature of the technique enables causality to be identified across numerous systemic levels (Leveson, 2004).
- The method can be, and has been, utilized in numerous domains.
- STAMP includes both a taxonomy of possible failures and a control structure template to guide the analyst in the identification of causal factors.
- It is a systemic method and as such is supported by a wealth of contemporary HF research promoting the systems approach.

6.3.6 Disadvantages

- The analysis is resource-intensive, especially with respect to time (Braband et al., 2003).
- A significant amount of detailed data is required to conduct the comprehensive method.
- Previous research has highlighted the need to increase the level of guidance within the STAMP method (Qureshi, 2007).

6.3.7 Related Methods

Conducting a STAMP analysis requires the utilization of data collection techniques such as observations and interviews.

6.3.8 Approximate Training and Application Times

STAMP analysis is a time-consuming procedure (Braband et al., 2003) and involves high training times.

6.3.9 Reliability and Validity

There is currently no data available on the reliability or validity of STAMP.

6.3.10 Tools Needed

A STAMP analysis can be conducted by using a pen and paper, but it benefits from software drawing packages such as Microsoft Office Visio to create high-quality control structure diagrams.

6.3.11 Example

STAMP has been applied to numerous incidents occurring within complex sociotechnical systems, for which the reader is referred to Leveson (2002). Figure 6.3 represents an extract of the analysis conducted into the land use and transport integration of the road transport system in Queensland, Australia (Salmon et al., 2016). The aim of the analysis was to identify the range of actors and organizations within the system along with the key relationships that exist between them.

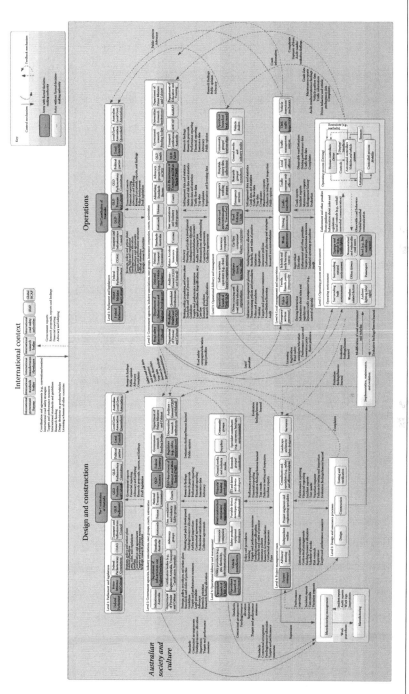

Figure 6.3 Land use and transport integration of the Queensland road transport system. (From Salmon, P. M. et al., *Accident Analysis & Prevention*, 96, 140–151, 2016a.)

6.3.12 Flowchart

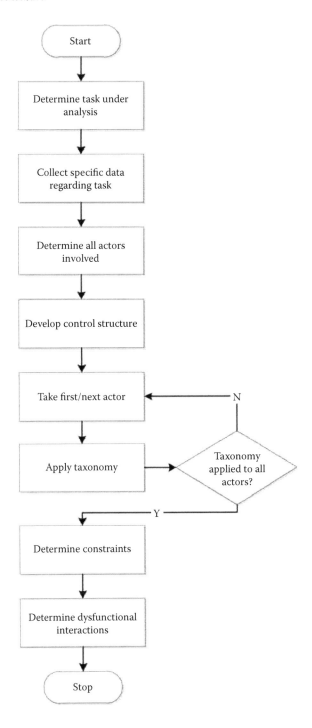

7

SITUATION AWARENESS ASSESSMENT

7.1 Introduction

Situation awareness (SA) is the term that is used within human factors (HF) circles to describe the level of awareness that individuals have of the situation that they are engaged in. It focuses on how they develop and maintain a sufficient understanding of "what is going on" (Endsley, 1995a) to achieve success in task performance. SA is a fundamental interaction of humans within our urban and regional systems and environments. In land use planning and urban design (LUP & UD), it is imperative that we acknowledge that people develop SA based on what we decide to include within the urban environment. It underpins how people relate to their surroundings, and indeed each other, while making decisions about the tasks and activities they perform. A more sophisticated approach to the understanding and assessment of SA by LUP & UD practitioners and researchers can only lead to the design of more efficient and human-orientated urban places.

While an essential commodity in the safety critical domains, SA is now recognized as a key consideration in system design and evaluation more broadly (e.g., Endsley, 2016; Salmon et al., 2009a; Stanton et al., 2017). Reflecting this, SA has been explored in a range of domains over the past 30 years, ranging from the military (e.g., Salmon et al., 2009) and transportation domains (e.g., Salmon, 2014a) to sport (James and Patrick, 2004; Neville and Salmon, 2016), health care and medicine (Hazlehurst et al., 2007), and the emergency services (e.g., Blandford and Wong, 2004). A contentious concept, various models of SA have been postulated, focusing either on the awareness held by individuals (e.g., Endsley, 2016; Smith and Hancock, 1995), teams (e.g., Endsley and Robertson, 2000; Salas et al., 1995), or sociotechnical systems (e.g., Salmon et al., 2009a; Stanton et al., 2017). As a corollary, various different approaches for assessing SA have been developed and applied in a range of domains.

7.2 Situation Awareness Theory

The concept first emerged as a topic of interest within the military aviation domain when it was identified as a critical asset for military aircraft crews during the World War I (Press, 1986; cited in Endsley, 1995a). Despite this, it did not begin to receive attention in academic circles until the late 1980s (Stanton et al., 2017), when SA-related research began to emerge within the aviation and air traffic control domains (e.g., Endsley, 1989, 1993).

Various definitions of SA are presented in the academic literature (e.g., Adams et al., 1995; Billings, 1995; Dominguez, 1994; Fracker, 1991; Sarter and Woods, 1991; Smith and Hancock, 1995; Stanton et al., 2006b; Taylor, 1990). Still, by far the most prominent is that offered by Endsley, who defines SA as a cognitive product (resulting from a separate process labeled situation assessment) comprising "the perception of the elements in the environment within a volume of time and space, the comprehension of their meaning, and the projection of their status in the near future" (Endsley, 1995a: 36). Given the increased presence of complex sociotechnical systems in which teams of humans together collaboratively, is Stanton et al.'s (2006b) systems theory–oriented definition, which asserts that SA represents "activated knowledge for a specific task, at a specific time within a system."

7.2.1 Individual Models of Situation Awareness

Inaugural SA models were, in the main, focused on how individual operators develop and maintain SA whilst undertaking activity within complex systems (e.g., Adams et al., 1995; Endsley, 1995a; Smith and Hancock, 1995). As such, these models primarily focus on the awareness held in the minds of individuals (operators), that is, SA as experienced in the mind of the person (Stanton et al., 2010b). Two of these models, in particular, stand out from the literature, one being by far the most popular, the other being, in the author's view at least, the most

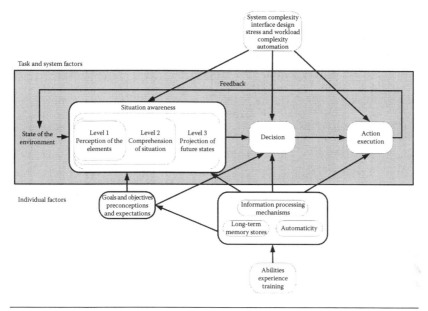

Figure 7.1 Endsley's three-level model of situation awareness. (Adapted from Endsley, M. R., *Human Factors*, 37, 32–64, 1995a.)

appropriate for describing the concept, albeit at an individual level. Endsley's three-level model (Endsley, 1995a, 2016) has undoubtedly received the most attention. The information processing–based model describes SA as an internally held cognitive product comprising three levels (Figure 7.1) that follows perception and leads to decision-making and action execution.

The first level involves perceiving the status, attributes, and dynamics of task-related elements in the surrounding environment. A range of factors influence the data perceived, including the task being performed, the individual's goals, experience, expectations, and also systemic factors such as interface design, level of complexity, and automation. To achieve level 2 SA, the individual interprets the level 1 data in such a way that they can then comprehend its relevance to their task and goals. The interpretation and comprehension of SA-related data is influenced by an individual's goals, expectations, experience in the form of mental models, and preconceptions regarding the situation. Level 3 SA involves prognosticating future system states. By using a combination of level 1 and 2 SA-related knowledge and experience in the form of mental models, individuals can forecast likely future states in the situation. For example, a transport planner can forecast, based on level 1 and 2-related information, that a particular point of congestion may occur under a certain circumstance. The planner can do this through perceiving elements such as the location of a roadway, its hierarchy, the number of vehicles, comprehending what the elements mean, and then comparing this with experience (in the form of mental models) to forecast what might happen next. Mental models are therefore used to facilitate the achievement of SA by directing attention to critical elements in the environment (level 1), integrating the elements to aid understanding of their meaning (level 2), and generating possible future states and events (level 3). According to the model, SA acquisition and maintenance are influenced by other factors, including individual factors (e.g., training and workload), task factors (e.g., complexity), and systemic factors (e.g., interface design) (Endsley, 1995a).

Smith and Hancock's ecological approach, based on Neisser's (1976) perceptual cycle model, takes a more holistic stance, viewing SA as a "generative process of knowledge creation and informed action taking" (Smith and Hancock, 1995: 138). According to the perceptual cycle model, our interaction with the world (termed explorations) is directed by internally held schemata. The outcome of interaction modifies the original schemata, which in turn directs further exploration. This process of directed interaction and modification continues in an infinite cyclical nature. By using this model, Smith and Hancock (1995) suggested that SA is neither resident in the world nor in the person but resides through the interaction of the person with the world. They describe SA as "externally, directed consciousness" that is an "invariant component in an adaptive cycle of knowledge, action and information" (1995: 138). In addition, they argue that the process of achieving and maintaining SA revolves around internally held schema, which contain information regarding certain situations. These schemata facilitate the anticipation of situational events, directing an individual's attention to cues in the environment, and directing his or her eventual course of action. An individual then conducts checks to confirm that the evolving situation conforms to his or her expectations. Any unexpected events serve to prompt further search and explanation, which in turn modifies the individual's existing model. The perceptual cycle model of SA is presented in Figure 7.2.

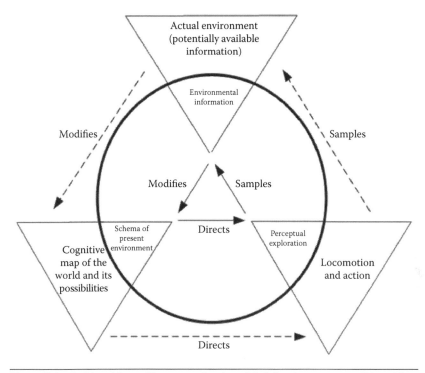

Figure 7.2 Smith and Hancock's perceptual cycle model of situation awareness. (Adapted from Smith, K. and Hancock, P. A., *Human Factors*, 37, 137–148, 1995.)

Smith and Hancock (1995) identified SA as a subset of the content of working memory in the mind of the individual (in one sense, it is a product). However, they emphasize that attention is externally directed rather than introspective (and thus is contextually linked and dynamic). Unlike the three-level model, which depicts SA as a product separate from the processes used to achieve it, SA is therefore viewed as both process and product. Smith and Hancock's complete model therefore views SA as more of a holistic process that influences the generation of situational representations. The model has sound underpinning theory (Neisser, 1976) and is complete in that it refers to the continuous cycle of SA acquisition and maintenance, including both the process (the continuous sampling of the environment) and the product (the continually updated schema) of SA. Their description also caters for the dynamic nature of SA and more clearly describes an individual's interaction with the world to achieve and maintain SA, whereas Endsley's model seems to place the individual as a passive information receiver. The model therefore considers the individual, the situation, and the interactions between the two.

7.2.2 Team Models of Situation Awareness

Of course, due to the nature of teamwork and the systems in which it takes place, team SA has to be more complex than individual SA. Various models of team SA have been proposed (e.g., Salas et al., 1995; Shu and Furuta, 2005; Stanton et al., 2006b, 2009). By reflecting the three-level model's widespread popularity, the most common approach to describing team SA has involved applying Endsley's model to team SA, along with the addition of the related but distinct concepts of *team* and *shared* SA (e.g., Endsley and Jones, 1997; Endsley and Robertson, 2000). Team SA reflects "the degree to which every team member possesses the SA required for his or her responsibilities" (Endsley, 2016), whereas shared SA refers to "the degree to which team members have the same SA on shared SA requirements" (Endsley and Jones, 1997). Endsley's approach to team SA therefore suggests that team members not only have distinct portions of SA but also overlapping or *shared* portions of SA. Successful team performance requires that individual team members have good SA on their specific elements and the same SA for shared SA elements (Endsley and Robertson, 2000).

Much like individual SA, the concept of team SA is plagued by contention. For example, many have expressed concern over the use of Endsley's individual operator three-level model to describe team SA (Artman and Garbis, 1998; Gorman et al., 2006; Patrick et al., 2006; Salmon et al., 2006, 2009a; Stanton et al., 2009, 2015; Shu and Furuta, 2005; Siemieniuch and Sinclair, 2006; Sonnenwald et al., 2004) and also regarding the relatively blunt characterization of shared SA (e.g., Salmon et al., 2009a; Stanton et al., 2009, 2010b, 2017). By putting aside the obvious concerns associated with using an individual information processing–based account of SA for describing team SA, the concept of shared SA remains ambiguous, and research in the area has questioned the notion that different human operators can *share* SA in a way that they understand a situation in exactly the same manner (e.g., Salmon et al., 2009a, 2016; Stanton et al., 2010b).

7.2.3 Models Accounting for Situation Awareness Held by Systems

Advances in the area have led to the description of SA as a social phenomenon, which is held by systems comprising human and technological agents. Known as "distributed situation awareness" (DSA) (Salmon et al., 2009a; Stanton et al., 2006b, 2015), this approach uses as its basis, distributed cognition-based accounts of system performance (e.g., Hutchins, 1995), which move the focus on cognition out of the heads of individual operators and on to the overall system consisting of human and technological agents; here, cognition transcends the boundaries of individual actors and *systemic* cognition is achieved by the transmission of representational states throughout the system (Hutchins, 1995).

SA was first discussed in this context by Artman and Garbis (1998), who, due to the flaws evident when applying individualistic models to complex sociotechnical systems, called for a systems perspective model on SA. They subsequently defined team SA as "the active construction of a model of a situation partly shared and partly distributed between two or more agents, from which one can anticipate important future states in the near future" (Artman and Garbis, 1998: 2). Following this, the foundations for a theory of DSA in complex systems, laid by Stanton et al. (2006), were built upon by Salmon et al. (2009a) who outlined a model of DSA, developed on the basis of applied research in a range of military and civilian command and control environments.

Briefly, Stanton et al. (2006), model is underpinned by four theoretical concepts: schema theory (e.g., Bartlett, 1932), genotype and phenotype schema, Neisser's (1976) perceptual cycle model of cognition, and, of course, Hutchin's (1995) distributed cognition approach. SA is viewed as an emergent property of collaborative systems, arising from the interactions between agents, both human and technological. According to Stanton et al. (2006b, 2009), a system's awareness comprises a network of information on which different components of the system have distinct views and ownership of information. Scaling the model down to individual team members, it is suggested that team member SA represents the state of their perceptual cycle (Neisser, 1976); individuals possess genotype schema that are triggered by the task-relevant nature of task performance, and during task performance, the phenotype schema comes to the fore. It is this task and schema-driven content of team member SA that brings the shared SA (Endsley and Robertson, 2000) notion into question. Rather than possess shared SA (which suggests that team members understand a situation or elements of a situation in the same manner), the model instead suggests that team members possess unique, but compatible, portions of awareness. Team members experience a situation in different ways, as defined by their own personal experience, goals, roles, tasks, training, skills, schemata, and so on. Compatible awareness is therefore the phenomenon that holds distributed systems together (Stanton et al., 2006b, 2009). Each team member has his or her own awareness related to the goals that he or she is working toward. This is not the same as other team members but is such that it enables him or her to work with adjacent team members. Although different team members may have access to the same information, differences in goals, roles, the tasks being performed, experience, and their schema mean that their resultant awareness of it is not shared; instead, the situation is viewed differently based on these factors. However, each team member's SA is compatible as it is different in content but is collectively required for the system to perform collaborative tasks optimally, from one team member to another, which bestows upon team SA an emergent behavior.

7.2.4 *Measuring Situation Awareness*

The popularity of the concept is such that various methods of measuring SA have been proposed and applied (for detailed reviews, see Salmon et al., 2006, 2009a). These SA measures can be broadly categorized into the following methodological groups: SA requirements analysis, freeze probe recall methods, real-time probe methods, posttrial subjective rating methods, observer rating methods, process indices, and team SA measures. SA requirements analysis forms the first step in an SA assessment effort and is used to identify what exactly it is that comprises SA in the scenario and environment in question. Endsley defines SA requirements as "those dynamic information needs associated with the major goals or sub-goals of the operator in performing his or her job" (2001: 8). According to Endsley, they concern not only the data that operators need but also how the data are integrated to address decisions. Matthews et al. (2004) highlighted the importance of conducting SA requirements analysis when developing reliable and valid SA metrics.

The current chapter focuses on the following SA assessment methods: the SA requirements analysis method (Endsley, 1993), the situation awareness global assessment technique (SAGAT: Endsley, 1995b), the situation awareness rating technique (SART: Taylor, 1990), and the propositional network method (Salmon et al., 2009a; Stanton et al., 2006b, 2009).

7.3 Situation Awareness Requirements Analysis

7.3.1 Background and Applications

SA requirements analysis is used to identify exactly what it is that comprises SA in the scenario and environment under analysis. According to Endsley (2001), SA requirements concern not only the data that operators need but also how the data are integrated to address decisions. Matthews et al. (2004) suggested that a fundamental step in developing reliable and valid SA metrics is to identify the SA requirements of a given task. Further, the authors point out that knowing what the SA requirements are for a given domain provides engineers, planners, and urban designers with a basis to develop optimal system designs that maximize human and urban system performance.

Endsley (1993) and Matthews et al. (2004) described a generic procedure for conducting SA requirements analyses that uses unstructured interviews with subject matter experts (SMEs), goal-directed task analysis, and questionnaires to determine the SA requirements for a particular task or system. Endsley's methodology focuses on SA requirements across the three levels of SA, specified in her information processing–based model of SA (level 1: perception of elements, level 2: comprehension of meaning, and level 3: projection of future states).

7.3.2 Domain of Application

The SA requirements analysis procedure is generic and has been applied in various domains, including the military (Bolstad et al., 2002; Matthews et al., 2004) and air traffic control (ATC) (Endsley, 1993).

7.3.3 Application in Land Use Planning and Urban Design

The SA requirements analysis is the first step in clearly identifying which urban development settings or scenarios may benefit most from an understanding of SA. For example, the SA requirements of different members of multidisciplinary teams in LUP & UD are a pertinent line of enquiry.

7.3.4 Procedure and Advice

Step 1: Define the Task or Scenario Under Analysis
The first step in an SA requirements analysis is to clearly define the task or scenario under investigation. It is recommended that the task be described clearly, including the different actors involved, the task goals, and the environment within which the task is to take place. An SA requirements analysis requires that the task be defined explicitly to ensure that the appropriate SA requirements are comprehensively assessed.

Step 2: Select Appropriate SMEs
The SA requirements analysis procedure is based upon eliciting SA-related knowledge from SMEs. Therefore, the analyst should next select a set of appropriate SMEs. The more experienced the SMEs are in the environment under analysis, the better, and the analyst should strive to use as many SMEs as possible to ensure comprehensiveness.

Step 3: Conduct SME Interviews
Once the scenario and environment under analysis is defined clearly and appropriate SMEs are identified, a series of unstructured interviews with the SMEs should be conducted. First, participants should be briefed on the topic of SA and the concept of SA requirements analysis. Following this, Endsley (1993) suggested that the SME should be asked to describe, in their own words, what they feel comprises *good* SA for the task in question. They should then be asked what they would want to know to achieve perfect SA. Finally, the SMEs should be asked to describe what each of the SA elements identified is used for within the setting or scenario under analysis, for example, decision-making, planning, and actions. Endsley

(1993) also suggested that once the interviewer has exhausted the SME's knowledge, he or she should offer his or her own suggestions regarding SA requirements and should discuss their relevance. It is recommended that each interview is recorded using either video- or audio-recording equipment. Following completion of the interviews, all data should be transcribed.

Step 4: Conduct Goal-Directed Task Analysis

Once the interview phase is complete, a goal-directed task analysis should be conducted for the task or scenario under investigation. Endsley (1993) prescribed her own goal-directed task analysis method; however, it is also possible to use hierarchical task analysis (HTA) (Annett et al., 1971) for this purpose as it focuses on goals and their decomposition. For this purpose, the HTA procedure presented in Chapter 3 should be used. Once the HTA is complete, the SA elements required for the completion of each step in the HTA should be added. This step is intended to ensure that the list of SA requirements identified during the interview phase is comprehensive. Upon completion, the task analysis output should be reviewed and refined using the SMEs utilized during the interview phase.

Step 5: Compile List of SA Requirements Identified

The outputs from the SME interview and goal-directed task analysis phases should then be used to compile a list of SA requirements for the different actors involved in the task under analysis.

Step 6: Rate SA Requirements

Endsley's method uses a rating system to sort the SA requirements identified on the basis of their importance. These should be compiled into a rating-type questionnaire, along with any other elements that the analyst feels are pertinent. Appropriate SMEs should then be asked to rate the criticality of each of the SA elements identified in relation to the task under analysis. Items should be rated as not important (1), somewhat important (2), or very important (3). The ratings provided should then be averaged across subjects for each item.

Step 7: Determine SA Requirements

Once the questionnaires have been collected and scored, the analyst should use them to determine the SA elements for the task or scenario under analysis. How this is done is dependent upon the analyst's judgment. It may be that the elements specified in the questionnaire are presented as SA requirements, along with a classification in terms of importance (e.g., not important, somewhat important, or very important).

Step 8: Create SA Requirements Specification

The final stage involves creating an SA requirements specification that can be used by other practitioners (e.g., architects, designers, or methods developers). The SA requirements should be listed for each actor involved in the task or scenario under analysis. Endsley (1993) and Matthews et al. (2004) demonstrated how the SA requirements can be categorized across the three levels of SA, as outlined by the three-level model; however, this may not be necessary, depending on the specification requirements. It is recommended that SA requirements should be listed in terms of what it is that needs to be known, what information is required, how this information is used (i.e., what the linked goals and decisions are), and what the relationships between the different pieces of information actually are, that is, how they are integrated and used by different actors. Once the SA requirements are identified for each actor in question, a list should be compiled, including tasks, SA elements, the relationships between them, and the goals and decisions associated with them.

7.3.5 *Advantages*

- SA requirements analysis provides a structured approach for identifying the SA requirements associated with a particular task or scenario.
- The output tells us exactly what it is that needs to be known by different actors during task performance.

- The output has many uses, including for developing SA measures or to inform the design of coaching and training interventions, procedures, or new technology.
- If conducted properly, it has the potential to be exhaustive.
- It is generic and can be used to identify the SA requirements associated with any task in any domain.

7.3.6 Disadvantages

- Due to the use of interviews and task analysis methods, the method is very time-consuming to apply.
- It requires a high level of access to multiple SMEs for the task under analysis.
- Identifying SA elements and the relationships between them requires significant skill on the part of the analyst involved.
- Analyses may become large and unwieldy for complex collaborative systems.
- Analysts require an in-depth understanding of the SA concept.
- It does not directly inform design.

7.3.7 Related Methods

The SA requirements analysis procedure outlined by Endsley (1993) was originally conceived as a way of identifying the SA elements to be tested using the SAGAT freeze probe recall method. The SA requirements analysis method itself uses interviews with SMEs and also goal-directed task analysis, which is similar to the HTA approach.

7.3.8 Approximate Training and Application Times

Provided that analysts have significant experience of the SA concept, interviews, and task-analysis methods, the training time for the SA requirements analysis method is low; however, for novice analysts new to the area and without experience in interview and task analysis methods, the time required is high. The application time for the SA requirements analysis method is high, including the conduct of interviews, transcription of interview data, conduct of task analysis for the task in question, the identification and rating of SA elements, and finally the compilation of SA requirements and the relationships between them.

7.3.9 Reliability and Validity

The reliability and validity of the SA requirements method is difficult to assess. As long as appropriate SMEs are used throughout the process, the validity should be high; however, the method's reliability may be questionable.

7.3.10 Tools Needed

At its most basic, the SA requirements analysis procedure can be conducted using pen and paper; however, to make the analysis as simple and as comprehensive as possible, it is recommended that video- and audio-recording equipment is used to record the interviews and that a computer with a word-processing package (such as Microsoft Word) and the Statistical Package for the Social Sciences is used during the design and analysis of the questionnaire. A drawing package such as Microsoft Visio is also useful when producing the task analysis and SA requirements analysis outputs.

7.3.11 Flowchart

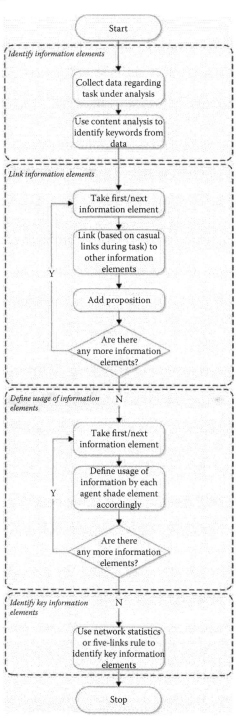

Level 1 SA requirements	Level 2 SA requirements	Level 3 SA requirements
Buildings Locations Construction Proximity Registers Mapping Architecture Ownership Facades Stories Purpose	Establish building address and footprints Identify the historical register Search the historical register Architectural features of buildings Building condition and current use Heights of individual buildings Year of construction Type of construction	Identify clusters of architectural representation Likelihood of buildings remaining Projected cues for any potential site designs

Figure 7.3 SA requirements extract for 6.4.1 identify surrounding buildings. Decisions: Where are the buildings? Are they historically significant? What is their size and shape? What height are they? What are they made of?

7.3.12 Example

For example, the purposes of the site analysis HTA were used to generate the SA requirements during a site analysis task. The example extract, Figure 7.3 focuses on the sub-goal decomposition for the task *Identify Surrounding Buildings*.

7.3.13 Recommended Text(s)

Endsley, M. R., Bolte, B., & Jones, D. G. (2003). *Designing for situation awareness: An approach to user-centred design*. London: Taylor & Francis Group.

Matthews, M. D., Strater, L. D., & Endsley, M. R. (2004). SA requirements for infantry platoon leaders. *Military Psychology*, 16, 149–161.

7.4 The Situation Awareness Global Assessment Technique

7.4.1 Background and Applications

SAGAT (Endsley, 1995b) is a freeze probe recall method that was developed to assess pilot SA based on the three levels of SA postulated by Endsley's information-processing model. It is a simulation-based measure and involves querying participants regarding their knowledge of SA elements during random freezes in a simulation of the task or scenario under analysis. During the freezes, all simulation screens and displays are blanked, and relevant SA queries for that point of the task or scenario are administered.

7.4.2 Domain of Application

SAGAT was originally developed for use in the military aviation domain; however, numerous variations of the method have since been applied in other domains, including an air-to-air tactical aircraft version (Endsley, 1990), an advanced bomber aircraft version (Endsley, 1989), and an ATC version (Endsley and Kiris, 1995). SAGAT-style approaches can be applied in any domain provided that the queries are developed on the basis of an SA requirements analysis for the domain and activities under investigation.

7.4.3 Application in Land Use Planning and Urban Design

SAGAT is useful for the SA assessment of existing urban environments or proposed urban environments. By using simulations or scenarios of either environment, SAGAT will allow designers' and decision maker's unique insights into the use and management of urban space. For example, it may assist in the identification of the SA of different types of users of public space; aged, young, disabled, frequent, or first-time visitors to a space. Further, it may assist in the evaluation of impact from the insertion of various technologies or urban-design configurations in an environment. Insights may afford better designs for safety, place attachment, and wayfinding.

7.4.4 Procedure and Advice

Step 1: Define the aims of the analysis

First, the aims of the analysis should be clearly defined as this affects the scenarios used and the types of SAGAT queries administered. For example, the aims of the analysis may be to evaluate the impact that a new performance aid, technological device, or urban design has on SA during task performance, or it may be to compare novice and expert performer SA during a particular task.

Step 2: Define the task or scenario under analysis

The next step involves clearly defining the task or scenario under analysis. It is recommended that the task be described clearly, including the different actors involved, the task goals, and the environment within which the task is to take place.

Step 3: Conduct SA requirements analysis and generate SAGAT queries

To support query development, an SA requirements analysis is required for the activity or system under investigation. The SA requirements analysis output is then used to inform the development of appropriate SAGAT queries. Jones and Kaber (2005) highlighted the importance of this phase, suggesting that the foundation of successful SAGAT data collection efforts rests solely on the efficacy of the queries used. The queries generated should cover the three levels of SA as prescribed by Endsley's model (i.e., perception, comprehension, and projection). Jones and Kaber stress that the wording of the queries should be compatible with the operator's frame of reference and appropriate to the language typically used in the domain under analysis.

Step 4: Brief participants

Once appropriate participants have been recruited on the basis of the analysis requirements, the data collection phase can begin. First, however, it is important to brief the participants involved. This should include an introduction to the area of SA and a description and demonstration of the SAGAT methodology. At this stage, participants should also be briefed on what the aims of the study are and what is required of them as participants.

Step 5: Conduct pilot run(s)

Before the data collection process proper begins, it is recommended that pilot runs of the SAGAT data collection procedure are undertaken. A number of small test scenarios, incorporating multiple SAGAT freezes and query administrations, should be used to iron out any problems with the data collection procedure, and the participants should be encouraged to ask any questions. Once the participant is familiar with the procedure and is comfortable with his or her role, the *real* data collection process can begin.

Step 6: Begin SAGAT data collection

Next, the SAGAT data collection phase can begin. The experimenter should initiate this by instructing the participant(s) to undertake the task under analysis.

Step 7: Freeze the simulation

SAGAT works by temporarily freezing the simulation at predetermined random points and blanking all displays or interfaces. Jones and Kaber (2005) offered the following guidelines for task freezes:

- The timing of freezes should be randomly determined.
- SAGAT freezes should not occur within the first 3 to 5 min of the trial.
- SAGAT freezes should not occur within 1 min of each other.
- Multiple SAGAT freezes should be used.

Step 8: Administer SAGAT queries

Once the simulation is frozen at the appropriate point, the analyst should probe the participant's SA using the predefined SA queries. These queries are designed to allow the analyst to gain a measure of the participant's knowledge of the situation at that exact point in time and should be directly related to the participant's SA at the point of the freeze. A computer programmed with the SA queries is normally used to administer the queries; however, queries can also be administered using pen and paper. To stop any overloading of the participants, not all SA queries are administrated in any one freeze, and only a randomly selected portion of the SA queries is administrated at any one time. Jones and Kaber (2005) recommended that no outside information should be available to the participants during query administration. For evaluation purposes, the correct answers to the queries should also be recorded; this can be done automatically by sophisticated computers/simulators or manually by an analyst. Once all queries are completed, the simulation should resume from the exact point at which it was frozen (Jones and Kaber, 2005). Steps 7 and 8 are repeated throughout the task until sufficient data are obtained.

Step 9: Query response evaluation and SAGAT score calculation

Upon completion of the simulator trial, participant's query responses are compared with what was actually or expected to happen in the situation at the time of the query administration. To achieve this, query responses are compared with the data recorded by the simulation computers or analysts involved. Endsley (1995b) suggested that this comparison of the real and perceived situation provides an objective measure of participant's SA. Typically, responses are scored as either correct (1) or incorrect (0), and a SAGAT score is calculated for each participant, including an overall score and scores for each of the three SA levels. Additional measures or variations on the SAGAT score can be taken depending on study requirements, such as time taken to answer queries.

Step 10: Analyze SAGAT data

SAGAT data are typically analyzed across conditions (e.g., trial 1 versus trial 2) and the three SA levels specified by Endsley's three-level model. This allows query responses to be compared across conditions and also levels of SA to be compared across participants.

7.4.5 Advantages

- SAGAT provides an online measure of SA, removing the problems associated with collecting subjective SA data (e.g., a correlation between SA ratings and task performance).
- Online data collection avoids the problems associated with collecting SA data posttask, such as memory degradation and forgetting low SA periods of the task.
- SA scores can be viewed in total and also across the three levels specified in Endsley's model. Further, the specification of SA scores across Endsley's three levels is useful for designers and easy to understand.
- SAGAT is the most popular approach for measuring SA and has the most validation evidence associated with it (Jones and Endsley, 2000, Durso et al., 1998, Endsley and Garland, 2000).

- Evidence suggests that SAGAT is a valid metric of SA (Jones and Kaber, 2005).
- The method is generic and can be applied in any domain.

7.4.6 Disadvantages

- Various preparatory activities are required, including the conduct of SA requirements analysis and the generation of numerous SAGAT queries.
- The total application time for the whole procedure (i.e., including SA requirements analysis and query development) can be high.
- By using the SAGAT method typically requires expensive high-fidelity simulators and computers.
- The use of task freezes and online queries is highly intrusive to performance on the primary task.
- It cannot be applied during real-world and/or collaborative tasks.
- It does not account for distributed cognition or distributed SA theory (Salmon et al., 2009a). For example, in a joint cognitive system, it may be that operators do not need to be aware of certain elements as they are held by displays and devices. In this case, SAGAT would score participant's SA as low even though this may not in fact be the case.
- It is based upon the three-level model of SA (Endsley, 1995a), which has various flaws (Salmon et al., 2008a).
- Participants may be directed to elements of the task that they are unaware of.
- To use the approach, one has to be able to determine what SA consists of a priori. This might be particularly difficult, if not impossible, for some scenarios.

7.4.7 Related Methods

SAGAT queries are generated based on an initial SA requirements analysis conducted for the task in question. Various versions of SAGAT have been applied, including an air-to-air tactical aircraft version (Endsley, 1990), an advanced bomber aircraft version (Endsley, 1989), and an ATC version (Endsley and Kiris, 1995). Situation awareness of en-route air traffic controllers in the context of automation (SALSA).

7.4.8 Approximate Training and Application Times

The training time for the SAGAT approach is low; however, if analysts require training in the SA requirements analysis procedure, then the training time incurred will increase significantly. The application time for the overall SAGAT procedure, including the conduct of an SA requirements analysis and the development of SAGAT queries, is typically high. The actual data collection process of administering queries and gathering responses requires relatively little time, although this is dependent upon the task under analysis.

7.4.9 Reliability and Validity

There is considerable validation evidence for the SAGAT approach presented in the literature. Jones and Kaber (2005) pointed out that numerous studies have been undertaken to assess the validity of the SAGAT, and the evidence suggests that the method is a valid metric of SA. Endsley (2000) reported that SAGAT has been shown to have a high degree of validity and reliability for measuring SA. Fracker (1991), however, reported low reliability for SAGAT when measuring participant's knowledge of aircraft location. Regarding validity, Endsley et al. (2000) reported a good level of sensitivity for SAGAT, but not for real-time probes (online queries with no freeze) and subjective SA measures.

7.4.10 Tools Needed

Typically, a high-fidelity simulation of the task or scenario under analysis and computers with the ability to generate and score SAGAT queries are required. The simulation and computer used should possess the ability to randomly blank all operator displays and *window* displays, randomly administer relevant SA queries, and calculate participant's SA scores.

7.4.11 Example

Based on the example of a *wayfinding* task, a series of SAGAT *freeze* probes for a pedestrian walking situation were developed. Level 1, 2, and 3 SA SAGAT probes for the task *walk on the sidewalk* are presented in Table 7.1.

Table 7.1   SAGAT Queries for the *Wayfinding Task Walk on the Sidewalk*

LEVEL 1 QUERIES

- Where are you going?
- What time is it?
- At what distance is your destination?
- Where are you?
- Where is the sidewalk?
- Where is the vehicle traffic and driveways?
- Where is the pedestrian traffic?
- Where are the obstacles?
- Where is the signage?
- Where is the free space on the sidewalk?
- Have you been given any instructions?
- What are the current weather conditions?

LEVEL 2 QUERIES

- What directional opportunities are available at the present moment?
- Can you see your destination?
- What obstacles are in the way?
- What type of traffic is there?
- Which of the sidewalk routes has free space?
- What are your instructions?
- What does the signage say?
- Have you got clear visibility?
- What are you looking at?

LEVEL 3 QUERIES

- Will the free space remain on the chosen route?
- Will any traffic slow you down?
- Which obstacles will stop you from walking?
- Will you be able to continue to walk?

7.4.12 Flowchart

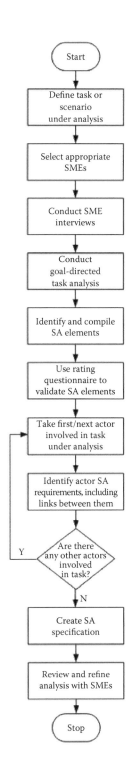

7.5 The Situation Awareness Rating Technique

7.5.1 Background and Applications

The SART (Taylor, 1990) is a posttrial subjective rating method, which uses the following 10 dimensions to measure operator SA: familiarity of the situation, focusing of attention, information quantity, information quality, instability of the situation, concentration of attention, complexity of the situation, variability of the situation, arousal, and spare mental capacity. SART is typically administered posttrial and involves participants subjectively rating each dimension on a seven-point rating scale (1 = low, 7 = high) based on their own perceived performance during the task under analysis. The ratings are then combined to calculate a measure of participant's SA. A quicker version of the SART approach also exists, known as the 3D SART. The 3D SART uses the 10 dimensions described earlier grouped into the following three dimensions:

1. *Demands on attentional resources*—A combination of complexity, variability, and instability of the situation
2. *Supply of attentional resources*—A combination of arousal, focusing of attention, spare mental capacity, and concentration of attention
3. *Understanding of the situation*—A combination of information quantity, information quality, and familiarity of the situation

7.5.2 Domain of Application

The SART approach was originally developed for use in the military aviation domain; however, the dimensions used are generic, and it has since been applied in various domains to assess operator SA, including command and control (Salmon et al., 2009a), and road transport (Walker et al., 2008).

7.5.3 Application in Land Use Planning and Urban Design

The SART approach could be applied in the LUP & UD context to provide a means by which users can rate urban spaces and provide designers and decision-makers with an assessment of their SA. Further, SART may be used to provide individual and team practitioner insights into the multidisciplinary assessment of land use planning scenarios or urban design concepts.

7.5.4 Procedure and Advice

Step 1: Define the tasks under analysis

The first step involves defining the tasks that are to be subjected to analysis. The type of tasks analyzed is dependent
upon the focus of the analysis. For example, when assessing
the effects on operator or user SA caused by a novel design, it
is useful to analyze as representative, a set of tasks as possible.

Step 2: Select participants

Once the task (or tasks) under analysis is clearly defined, it is
useful to select the participants who are to be involved in the
analysis. This may not always be necessary, and it may suffice
to select participants randomly on the day; however, if SA is
being compared across experience or ability levels, then effort
is required to select the appropriate participants.

Step 3: Brief participants

Before the task (or tasks) under analysis is performed, all the
participants involved should be briefed regarding the purpose
of the study, the concept of SA, and the SART method. It may
useful at this stage to take the participants through an example
SART analysis so that they understand how the method works
and what is required of them as participants. Explanation of
the different SART dimensions should also be provided.

Step 4: Conduct pilot run

It is recommended that participants take part in a pilot run
of the SART data collection procedure. A number of small
test scenarios should be used to iron out any problems with
the data collection procedure, and the participants should be
encouraged to ask any questions. Once the participants are
familiar with the procedure and are comfortable with their
role, the data collection process can begin.

Step 5: Performance of task

The next step involves the performance of the task or scenario
under analysis. Participants should be asked to perform the
task as normal or as they would usually do it.

Step 6: Complete SART questionnaires

Once the task is completed, participants should be given
a SART pro forma and asked to provide ratings for each

dimension based on how they felt during task performance. The participants are permitted to ask questions to clarify the dimensions; however, the participants' ratings should not be influenced in any way by external sources. To reduce the correlation between SA ratings and performance, no performance feedback should be given until after the participants have completed the self-rating process.

Step 7: Calculate participants' SART scores

The final step in the SART analysis involves calculating each participant's SA score. When using SART, participant's SA is calculated using the following formula: $SA = U - (D - S)$ where U = summed understanding, D = summed demand, S = summed supply. Typically, for each participant, an overall SART score is calculated along with total scores for the following three dimensions: understanding, demand, and supply.

7.5.5 *Advantages*

- SART is quick and simple to apply.
- It requires little training, both for analysts and participants.
- The data obtained are easily and quickly analyzed.
- It provides a low-cost approach to assessing participant's SA.
- Its dimensions are generic, so it can be applied in any domain.
- It is nonintrusive to task performance.
- It is a widely used method and has been applied in a range of different domains.
- It provides a quantitative assessment of SA.

7.5.6 *Disadvantages*

- SART suffers from a host of problems associated with collective subjective SA ratings, including a correlation between performance and SA ratings, and questions regarding whether or not participants can accurately rate their own awareness (e.g., how can one be aware that they are not aware?).
- It suffers from a host of problems associated with collecting SA data posttask, including memory degradation and poor recall, a correlation of SA ratings with performance, and also participants forgetting low SA portions of the task.

- Its dimensions are not representative of SA. Upon closer inspection, the dimensions are more representative of workload than anything else.
- It has performed poorly in a number of SA methodology comparison studies (e.g., Endsley et al., 2000; Endsley et al., 1998; Salmon et al., 2009b).
- The method is dated.

7.5.7 Related Methods

SART is one of many questionnaire-based, subjective-rating SA measurement approaches. Other such methods include the situation awareness rating scale (SARS: Waag and Houck, 1994), the crew awareness rating scale (CARS: McGuinness and Foy, 2000), and the situation awareness subjective workload dominance (SA-SWORD) method (Vidulich and Hughes, 1991).

7.5.8 Approximate Training and Application Times

The training and application times associated with the SART method are very low. As it is a self-rating questionnaire, there is very little training involved. In our experience, the SART questionnaire takes no longer than 5 min to complete, and it is possible to set up programs that autocalculate SART scores based on raw-data entry.

7.5.9 Reliability and Validity

Along with SAGAT, SART is the most widely used and tested measure of SA (Endsley and Garland, 2000); however, it has performed poorly in a number of validation and methods comparison studies (e.g., Endsley et al., 2000; Endsley et al., 1998; Salmon et al., 2009b). The construct validity of SART is limited, and many have raised concerns regarding the degree to which its dimensions are actually representative of SA (e.g., Endsley, 1995b; Salmon et al., 2009b; Uhlarik and Comerford, 2002).

7.5.10 Tools Needed

SART is applied using pen and paper only; however, it can also be administered using Microsoft Excel, which can also be used to automate the SART score calculation process.

7.5.11 Example

The SART pro forma is presented in Figure 7.4.

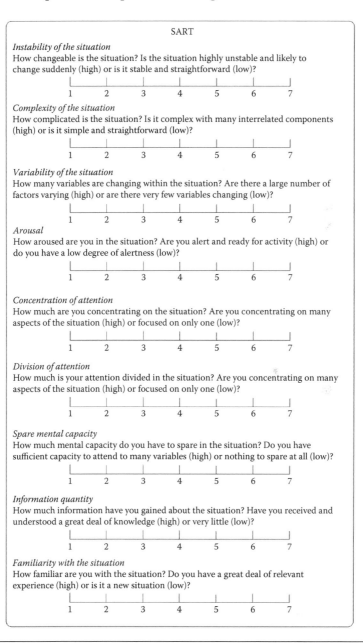

Figure 7.4 SART rating scale. (Modified from Taylor, R. M., Situational awareness rating technique (SART): The development of a tool for aircrew systems design, in *Situational Awareness in Aerospace Operations* (*AGARD-CP-478*), Neuilly Sur Seine, NATO-AGARD, Paris, France, pp. 3/1–3/17, 1990.)

7.5.12 Flowchart

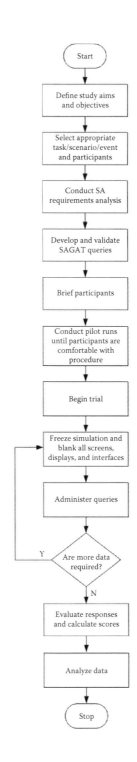

7.5.13 Recommended Text(s)

Taylor, R. M. (1990). Situational Awareness Rating Technique (SART): The development of a tool for aircrew systems design. In Situational Awareness in Aerospace Operations (AGARD-CP-478), pp. 3/1–3/17, Paris: Neuilly Sur Seine, NATO-AGARD.

7.6 Propositional Networks

7.6.1 Background and Applications

The propositional network methodology (Stanton et al., 2006b; Salmon et al., 2009a) is used for modeling DSA in collaborative systems. As existing measures such as SAGAT and SART focus exclusively on the levels of SA held by individual operators, the propositional network approach was proposed as a way of modeling the SA held by sociotechnical systems comprising both human and technological agents. Propositional networks use networks of linked information elements to depict the information underlying a system's awareness, the relationships between the different pieces of information, and also how each component of the system is using each piece of information. Salmon et al. (2009a) argued that, in addition to the information underpinning SA, it is the links between the different pieces of information that are more important in terms of understanding the concept. When using propositional networks, DSA is therefore represented as information elements (or concepts) and the relationships between them, which relate to the assumption that knowledge comprises concepts and the relationships between them (Shadbolt and Burton, 1995). Anderson (1983) first proposed the use of propositional networks to describe activation in memory. They are similar to semantic networks in that they contain linked nodes; however, they differ from semantic networks in two ways (Stanton et al., 2006b). First, rather than being added to the network randomly, the words are instead added through the definition of propositions. A proposition in this sense represents a basic statement. Second, the links between the words are labeled to define the relationships between the propositions, that is, elephant *has* tail and mouse *is* rodent. Following Crandall et al. (2006), a propositional network about propositional networks is presented in Figure 7.5.

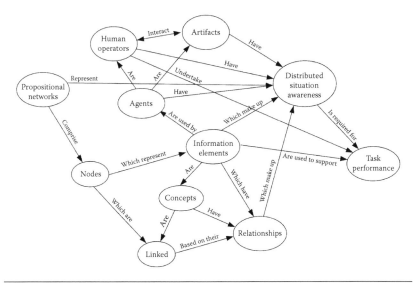

Figure 7.5 Propositional network diagram about propositional networks. (From Salmon, P. M. et al., *Distributed Situation Awareness: Advances in Theory, Measurement and Application to Teamwork*, Ashgate, Aldershot, UK, 2009.)

7.6.2 Domain of Application

The approach was originally applied for modeling DSA in command and control scenarios in the military (e.g., Stanton et al., 2006b; Salmon et al., 2009a) and civilian (e.g., Salmon et al., 2008b, 2014) domains; however, the method is generic and can be applied in any domain. Propositional networks have since been applied in a range of fields, including road transport (Salmon, 2015; Walker et al., 2009a), naval warfare (Stanton et al., 2006b), land warfare (Salmon et al., 2009a), railway maintenance operations (Walker et al., 2006), and military aviation (Stewart et al., 2008).

7.6.3 Application in Land Use Planning and Urban Design

The identification of DSA in urban settings is critical where we rely on knowledge from artifacts (e.g., signs, public art, and materials) and cognition to navigate a space. Propositional networks can reveal the intuition that is required and expected with the design of an urban environment, and also the impact of technology within an urban setting.

7.6.4 Procedure and Advice

Step 1: Define the aims of the analysis

First, the aims of the analysis should be clearly defined as this affects the scenarios used and the propositional networks developed. For example, the aims of the analysis may be to evaluate DSA during task performance or to evaluate the impact that a new training intervention, assessment tool, procedure, or technological device has on DSA during task performance.

Step 2: Define task or scenario under analysis

The next step involves clearly defining the task or scenario under analysis. It is recommended that the task is described clearly, including the different actors involved, the task goals, and the environment within which the task is to take place. The HTA method is useful for this purpose.

Step 3: Collect data regarding the task or scenario under analysis

Propositional networks can be constructed from a variety of data sources, depending on whether DSA is being modeled (in terms of what it should or could comprise) or assessed (in terms of what it did comprise). These include observational study and/or verbal transcript data, critical decision method data, HTA data, or data derived from work-related artifacts such as standard operating instructions, user manuals, standard operating procedures, and conditions of normal use. Data should be collected regarding the task based on the opportunities available, although it is recommended that, as a minimum, the task in question is observed and verbal transcript recordings are made.

Step 4: Define concepts and relationships between them

It is normally useful to identify distinct task phases. This allows propositional networks to be developed for each phase, which is useful for depicting the dynamic and changing nature of DSA throughout a task or scenario. To construct propositional networks, first the concepts need to be defined, followed by the relationships between them. For the purposes of DSA assessments, the term *information elements* is used to refer to concepts. To identify the information elements related to the task under analysis, a simple content analysis is performed on the input data and keywords are extracted. These keywords represent the information elements, which are then linked on the basis of their causal links during the activities in question (e.g., pedestrian *has* right of way, plants *buffer* roadway). Links are represented by directional arrows and should be overlaid with the linking proposition.

The output of this process is a network of linked information elements; the network contains all of the information that is used by the different actors and artifacts during task performance and thus represents the system's awareness, or what the system *needed to know* to successfully undertake task performance.

Step 5: Define information element usage

Information element usage is normally represented via shading of the different nodes within the network based on their usage by different actors during task or scenario performance. During this step, the analyst identifies which information elements the different agents, including both human and technological agents used. This can be done in a variety of ways, including by further analyzing input data (e.g., observational transcripts, verbal transcripts, and HTA) and by holding discussions with those involved or relevant SMEs.

Step 6: Review and refine network

Constructing propositional networks is a highly iterative process that normally requires numerous reviews and reiterations. It is recommended that once a draft network is created, it is subject to at least three reviews. It is normally useful to involve domain SMEs or the participants who performed the task in this process. The review normally involves checking the information elements and the links between them and also the usage classification. Reiterations to the networks normally include the addition of new information elements and links, the revision of existing information elements and links, and also the modification of the information element usage based on SMEs' opinion.

Step 7: Analyze networks mathematically

Depending on the aims and requirements of the analysis, it may also be pertinent to analyze the propositional networks mathematically using social network statistics. For example, in the past, we have used sociometric status and centrality calculations to identify the key information elements within propositional networks. Sociometric status provides a measure of how *busy* a node is relative to the total number of nodes present within the network under analysis (Houghton et al., 2006). In this case, sociometric status gives an indication of the relative prominence of information elements based on their links to other information elements in the network. Centrality is also a metric of the standing of a node within a network (Houghton et al., 2006), but here, this standing is in terms of its *distance* from all other nodes in the network. A central node is one that is close to all other nodes in the network and a message conveyed from that node to an arbitrarily selected other node in the network would, on average, arrive via the least number of relaying hops (Houghton et al., 2006). Key information elements are defined as those that have salience for each scenario phase, salience being defined as those information elements that act as hubs to other knowledge elements. Those information elements with a sociometric status value above the mean sociometric status value and a centrality score above the mean centrality value are identified as key information elements.

7.6.5 *Advantages*

- Propositional networks depict the information elements underlying a system's DSA and the relationships between them.
- In addition to modeling the system's awareness, they also depict the awareness of individuals and subteams working within the system.
- The networks can be analyzed mathematically to identify the key pieces of information underlying a system's awareness.
- Unlike other SA measurement methods, they consider the mapping between the information elements underlying SA.
- The propositional network procedure avoids some of the flaws typically associated with SA measurement methods, including intrusiveness, high levels of preparatory work (e.g., SA requirements analysis and development of probes), and the

problems associated with collecting subjective SA data and SA data posttrial.

- The outputs can be used to inform training, system, device, and interface design, and evaluation.
- Easy to learn and use.
- Software support is available via the workload, error, situational awareness, time and teamwork (WESTT) software tool (Houghton et al., 2008).

7.6.6 Disadvantages

- Constructing propositional networks for complex tasks can be very time-consuming and laborious.
- It is difficult to present larger networks within articles, reports, and/or presentations.
- No numerical value is assigned to the level of SA achieved by the system in question.
- Those offer more of a modeling approach than a measure, although SA failures can be represented.
- The initial data collection phase may involve a series of activities and often adds considerable time to the analysis.
- Many find the departure from viewing SA in the heads of individual operators (i.e., what operators know) to viewing SA as a systemic property that resides in the interactions between actors and between actors and artifacts (i.e., what the system knows) a difficult one to grasp.
- The reliability of the method is questionable, particularly when being used by inexperienced analysts.

7.6.7 Related Methods

Propositional networks are similar to other network-based knowledge representation methods such as semantic networks (Eysenck and Keane, 1990) and concept maps (Crandall et al., 2006). The data collection phase typically utilizes a range of approaches, including observational study, critical decision method interviews, verbal protocol analysis, and HTA. The networks can also be analyzed using metrics derived from social network analysis methods.

7.6.8 Approximate Training and Application Times

Provided the analysts involved have some understanding of DSA theory, the training time required for the propositional network method is low; our experiences suggest that around one or two hours of training is required. Following training, however, considerable practice is required before analysts become proficient in the method. The application time is typically high, although it can be low if the task is simplistic and short.

7.6.9 Reliability and Validity

The content analysis procedure should ease some reliability concerns; however, the links between concepts are made on the basis of the analyst's subjective judgment and so the reliability of the method may be limited, particularly when being used by inexperienced analysts. The validity of the method is difficult to assess, although our experiences suggest that validity is high, particularly when appropriate SMEs are involved in the process.

7.6.10 Tools Needed

On a simple level, propositional networks can be conducted using pen and paper; however, video- and audio-recording devices are typically used during the data collection phase and a drawing package such as Microsoft Visio is used to construct the propositional networks. Houghton et al. (2008) describe the WESTT software tool that contains a propositional network construction module that autobuilds propositional networks based on text data entry.

7.6.11 Example

The propositional network in Figure 7.6 represents the collaborative system of the design of a children's playground environment. It shows the networks of linked information depicting the underlying system awareness of this playground environment. It shows the relationships between the different pieces of information and also how each component of the system is using each piece of information.

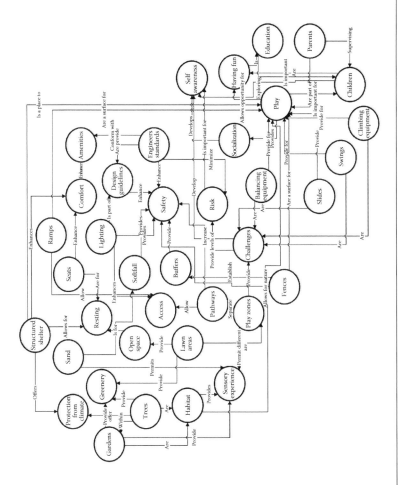

Figure 7.6 Propositional network of playground design. (Adapted from Missen, L. et al., A sociotechnical systems analysis approach to playground design, in *Contemporary Ergonomics 2017, Proceedings of the Chartered Institute of Ergonomics and Human Factors Annual Conference*, April, Daventry, UK, 2017.)

7.6.12 Flowchart

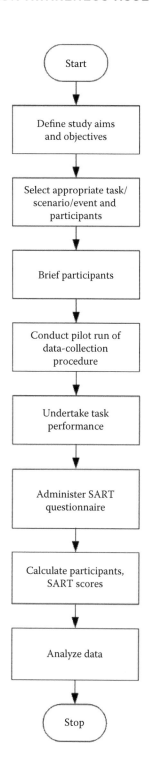

7.6.13 Recommended Text(s)

Salmon, P. M., Stanton, N. A., Walker, G. H., & Jenkins, D. P. (2009). *Distributed situation awareness: Advances in theory, modelling and application to teamwork*. Aldershot: Ashgate.

8

MENTAL WORKLOAD
ASSESSMENT METHODS

8.1 Introduction

The assessment of mental workload (MWL) is of crucial importance to the design and evaluation of built environments. The increasing complexity of urban and regional settings and the greater degrees of embedded intelligence mirrored with the increased role of technology has led to a greater level of demand being imposed on users of urban systems. MWL assessment recognizes that individuals possess a finite attentional capacity, and these attentional resources are allocated to relevant tasks. MWL represents the proportion of resources demanded by a task or set of tasks, and an excessive demand typically results in performance degradation (Young et al., 2015).

There has been much debate as to the nature of MWL, with countless attempts at providing a definition. Rather than reviewing these (often competing) definitions, we opt for the approach proposed by Megaw (2005), which is to consider MWL in terms of a framework of interacting stressors on an individual (Figure 8.1). The arrows indicate the direction of effects within this framework and imply that when we measure MWL, we are examining the impact of a whole host of factors on both performance and response. Clearly, this means that we are facing a multidimensional problem that is not likely to be amenable to single measures.

The construct of MWL has been investigated in a wide variety of domains, including aviation, air traffic control (ATC), military operations, driving, and control room operation to name but a few (Young et al., 2015). The assessment or measurement of MWL is used throughout the design life cycle to inform system and task design and to provide an evaluation of MWL imposed by existing operational systems and procedures. MWL assessment is also used to evaluate

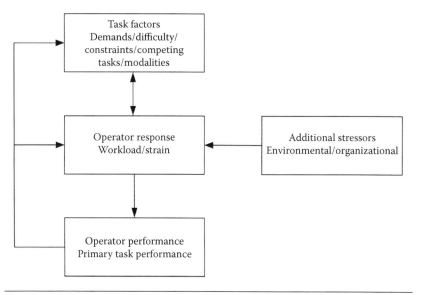

Figure 8.1   Framework of interacting stressors impacting workload. (Adapted from Megaw, T., The definition and measurement of mental workload, in J. R. Wilson and N. Corlett [Eds.], *Evaluation of Human Work*, pp. 525–553, CRC Press, Boca Raton, FL, 2005.)

the workload imposed during the operation of existing systems. There are a number of different MWL assessment procedures available to the researcher and practitioner. Traditionally, using a single approach to measure operator MWL has proved inadequate, and as a result, a combination of the methods available is typically used.

The assessment MWL often requires the use of a range of MWL assessment techniques, including primary task performance measures, secondary task performance measures (reaction times, embedded tasks), physiological measures (heart rate [HR], heart rate variability [HRV]), and subjective rating techniques (subjective workload assessment technique [SWAT], the national aeronautics and space administration [NASA] task load index [TLX]). This chapter will outline three subjective rating MWL assessment techniques. Each of these is valuable for the users of urban environments or indeed to explore practitioner land use planning processes and provide ratings of their perceived MWL during task performance. Subjective rating-assessment techniques are also attractive due to their ease and speed of application, and also the low cost involved. They are also unobtrusive to primary task performance and can be used in the field in *real-world* urban settings. MWL assessment may assist in land use planning and urban design (LUP & UD) to explore the inefficient use of, or indeed abuse of, valuable urban resources. If the users of an urban environment find it confusing, difficult to navigate, or have subjective feelings of poor safety, it will not be well used (Stevens, 2016). The following HF methods offer ways in which LUP & UD can better understand the MWL requirements of existing urban environments.

The NASA-TLX (Hart and Staveland, 1988) is a multidimensional subjective rating tool that is used to derive an MWL rating based upon a weighted average of six workload subscale ratings. The six subscales are mental demand, physical demand, temporal demand, effort, performance, and frustration level. The TLX is the most commonly used subjective MWL assessment technique, and there have been a number of validation studies associated with this technique. The Bedford scale (Roscoe and Ellis, 1990) uses a hierarchical decision tree to assess spare capacity whilst performing a task. Participants simply follow the decision tree to gain a workload rating for the task under analysis. The instantaneous self-assessment (ISA) of workload technique involves participants self-rating their workload during a task (normally every 2 min) on a scale of 1 (low) to 5 (high).

8.2 The National Aeronautics and Space Administration Task Load Index

8.2.1 Background and Applications

NASA-TLX (Hart and Staveland, 1988) is a subjective workload assessment tool that is used to gather subjective ratings of operator MWL in complex human and technical systems, such as aircraft pilots, process control room operators, and command and control system commanders. It is a multidimensional rating tool that gives an overall workload rating based upon a weighted average of six workload subscale ratings.

The six subscales and their associated definitions are given in the following:

- *Mental demand*: How much mental demand and perceptual activity was required (e.g., thinking, deciding, calculating, remembering, looking, searching)? Was the task easy or demanding, simple or complex, exacting or forgiving?

- *Physical demand*: How much physical activity was required (e.g., pushing, pulling, turning, controlling, activating)? Was the task easy or demanding, slow or brisk, slack or strenuous, restful or laborious?
- *Temporal demand*: How much time pressure did you feel due to the rate or pace at which the tasks or task elements occurred? Was the pace slow and leisurely or rapid and frantic?
- *Effort*: How hard did you have to work (mentally and physically) to accomplish your level of performance?
- *Performance*: How successful do you think you were in accomplishing the goals of the task set by the analyst (or yourself)? How satisfied were you with your performance in accomplishing these goals?
- *Frustration level*: How insecure, discouraged, irritated, stressed, and annoyed versus secure, gratified, content, relaxed, and complacent did you feel during the task?

Each subscale is presented to the participants either during or after the experimental trial, and they are asked to rate their score based upon an interval scale divided into 20 intervals, ranging from low (1) to high (20). The NASA-TLX also employs a paired comparisons procedure, whereby participants select the scale from each pair that has the most effect on the workload during the task under analysis. A total of 15 pairwise combinations are presented to the participants. This procedure accounts for two potential sources of between-rater variability: differences in workload definition between the raters and also differences in the sources of workload between the tasks.

The NASA-TLX is the most commonly used subjective MWL assessment technique and has been applied in numerous settings, including nursing (Malekpour et al., 2014), civil and military aviation, off-road driving (Stevens et al., 2016), nuclear power plant control room operation, and ATC. The tool has its own website http://human-factors.arc.nasa.gov/groups/TLX in which both versions (paper and computer) are available along with a list of publications.

8.2.2 Domain of Application

Generic.

8.2.3 Application in Land Use Planning and Urban Design

NASA-TLX can be utilized for better understanding and exploring a variety of LUP & UD tasks, designs, and processes. When considering the need and aspiration for better self-containment (live, work, and play) of our urban environments, it is clear that the measurement of six subscales of this technique is imperative. Further, that NASA-TLX can provide, for the first time, LUP & UD practitioners and researchers with MWL assessments for the range of different users of these existing and proposed urban settings.

8.2.4 Procedure and Advice (Computerized Version)

Step 1: Define the Task Under Analysis

The first step in a NASA-TLX analysis (aside from the process of gaining access to any required environments and participants) is to define the task (or tasks) that is to be subjected to analysis. The types of task analyzed are dependent upon the focus of the analysis. For example, when assessing the effects on user workload caused by a novel design or a new process, it is useful to analyze, as representative, a set of tasks as possible. To analyze a full set of tasks will often be too time-consuming and labor-intensive, and so it is pertinent to use a set of tasks that utilize all aspects of the system under analysis.

Step 2: Conduct a Hierarchical Task Analysis (HTA) for the Task under Analysis

Once the task (or tasks) under analysis is defined clearly, an HTA should be conducted for each task. This allows the analyst and participants to understand the task fully.

Step 3: Select Participants

Once the task (or tasks) under analysis is clearly defined and described, it may be useful to select the participants that are to be involved in the analysis. This may not always be necessary, and it may suffice to simply select participants randomly on the day. However, if workload is being compared across rank or experience levels, then clearly effort is required to select the appropriate participants.

Step 4: Brief Participants

Before the task (or tasks) under analysis is performed, all the participants involved should be briefed regarding the purpose of the study and the NASA-TLX technique. It is recommended that participants are given a workshop on workload and workload assessment. It may also be useful at this stage to take the participants through an example NASA-TLX application so that they understand how the technique works and what is required of them as participants. It may even be pertinent to get the participants to perform a small task and then get them to complete a workload profile questionnaire. This would act as a *pilot run* of the procedure and would highlight any potential problems.

Step 5: Performance of Task under Analysis

Next, the participant should perform the task under analysis. The NASA-TLX Figure 8.2 can be administered during the trial or after the trial. It is recommended that the TLX is administered after the trial as online administration is intrusive to the primary task. If online administration is required, the TLX should be administered and completed verbally.

Step 6: Weighting Procedure

When the task under analysis is complete, the weighting procedure can begin. The WEIGHT software presents 15 pairwise comparisons of the six subscales (mental demand, physical demand, temporal demand, effort, performance, and frustration level) to the participants. The participants should be instructed to select, from each of the 15 pairs, the subscale that contributed the most to the workload of the task. The WEIGHT software then calculates the total number of times each subscale was selected by the participant. Each scale is then rated by the software based upon the number of times it is selected by the participant. This is done using a scale of 0 (not relevant) to 5 (more important than any other factor).

Step 7: NASA-TLX Rating Procedure

Participants should be presented with the interval scale for each of the TLX subscales (this is done via the RATING software). Participants are asked to give a rating for each subscale, between 1 (low) and 20 (high), in response to the

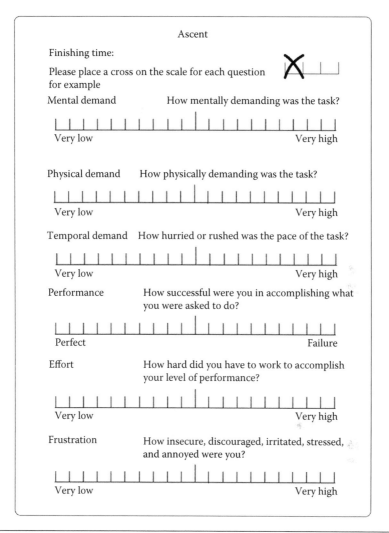

Figure 8.2 Example NASA-TLX pro forma. (From Salmon, P. M. et al., *Distributed Situation Awareness: Advances in Theory, Measurement and Application to Teamwork*, Ashgate, Aldershot, UK, 2009.)

associated subscale questions. The ratings provided are based entirely on the participant's subjective judgment.

Step 8: TLX Score Calculation

The TLX software is then used to compute an overall workload score. This is calculated by multiplying each rating by the weight given to that subscale by the participant. The sum of the weighted ratings for each task is then divided by 15 (the sum of weights). A workload score of between 0 and 100 is then provided for the task under analysis.

8.2.5 Advantages

- The NASA-TLX provides a quick and simple technique for estimating operator workload.
- Its subscales are generic, so the technique can be applied to any domain. In the past, it has been used in a number of different domains, such as aviation, ATC, command and control, nuclear reprocessing, petrochemical, and automotive domains.
- It has been tested thoroughly in the past and has also been the subject of a number of validation studies, for example, Hart and Staveland (1988).
- The provision of the TLX software package removes most of the work for the analyst, resulting in a very quick and simple procedure.
- For those without computers, it is also available in a pen-and-paper format (Vidulich and Tsang, 1985). It is probably the most widely used technique for estimating operator workload.
- It is a multidimensional approach to workload assessment.
- A number of studies have shown its superiority over SWAT (Hart and Staveland, 1988; Nygren, 1991).
- When administered posttrial, the approach is nonintrusive to primary task performance.
- According to Wierwille and Eggemeier (1993), it has demonstrated sensitivity to demand manipulations in numerous flight experiments.

8.2.6 Disadvantages

- When administered online, the TLX can be intrusive to primary task performance.
- When administered after the fact, participants may have forgotten high workload aspects of the task.
- Workload ratings may be correlated with task performance, for example, subjects who performed poorly on the primary task may rate their workload as very high and vice versa.
- The subscale weighting procedure is laborious and adds more time to the procedure.

8.2.7 Flowchart

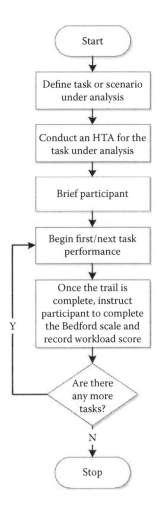

8.2.8 Example

A walking task by participants of different ages was undertaken as part of a study exploring the realities of the often used and arbitrary 400 m walkable catchment. Figure 8.2 represents an example of the NASA-TLX proforma. The collective data sheets of the pilot study (Figure 8.3) show the results of five participants undertaking a 400 m walking task adjacent to a busy urban roadway. Of interest, here are the frustration levels of the two *older* participants.

The real walkable catchment: What elements affect walkability amongst different age groups?

NASA-TLX (Cognitive load survey)

Walkability

Description of survey:

This NASA-TLX cognitive load survey is a workload assessment tool. This survey is used to address factors like mental demands, physical demands, temporal demands, own performance, effort, and frustration.

Instructions:

Below are a series of questions to be answered regarding the different demands, which may have been utilized whilst undertaking the walkability task at hand. Below the question is a scale with low and high indicators. Using the knowledge from the walk just undertaken answer each question by placing an "X" on the scale bar where you believe is an accurate representation of the load required in each section from the task just undertaken.

Mental demand

How much mental and perceptual activity was required (e.g., thinking, deciding, calculating, remembering, looking, searching)? Was the task easy or demanding, simple or complex, exacting or forgiving?

Low___X_____X__X_X_____X_____High

Physical demand

How much physical activity was required for example, pushing, pulling, turning, controlling, activation, and so on. Was the task easy or demanding, slow or brisk, slack or strenuous, restful or laborious?

Low__X_____X_X___X___X_____High

Temporal demand

How much time pressure did you feel due to the rate or pace at which the tasks or task elements occurred? Was the pace slow and leisurely or rapid and frantic?

Low_X__X_____X_X_____X_____High

Effort

How hard did you have to work (mentally and physically) to accomplish your level of performance?

Low__X_X_____X__X_X_____X_____High

Performance

How successful do you think you were in accomplishing the goals of the task set by the analyst (or yourself)? How satisfied were you with your performance in accomplishing these goals?

Low_____X_X__X__X_X_____High

Frustration level

How insecure, discouraged, irritated, stressed, and annoyed versus secure, gratified, content, relaxed, and complacent did you feel during the task?

Low__X____X____X_____X_____X__High

Figure 8.3 Pilot walkability study NASA-TLX work sheets. (From Pratt, J., *The Real Walkable Catchment: What Elements Affect Walkability amongst Different Age Groups? Unpublished Honors Theses*. Bachelor of Regional and Urban Planning [Honors], University of the Sunshine Coast, 2017.)

8.2.9 Related Methods

The NASA-TLX technique is one of a number of multidimensional subjective workload assessment techniques. Other multidimensional techniques include SWAT, the Bedford scale, DRA workload scales (DRAWS), and the MAlvern capacity estimate (MACE) technique. When conducting a NASA-TLX analysis, a task analysis (such as an HTA) of the task or scenario is often conducted. In addition, subjective workload assessment techniques are normally used in conjunction with other workload assessment techniques, such as primary and secondary task performance measures. To weight the subscales, the TLX uses a pairwise comparison weighting procedure.

8.2.10 Approximate Training Times and Application Times

The NASA-TLX technique is simple to use and quick to apply. The training times and application times are typically low.

8.2.11 Reliability and Validity

A number of validation studies concerning the NASA-TLX have been conducted (Hart and Staveland, 1988; Vidulich and Tsang, 1985, 1986). Vidulich and Tsang (1985, 1986) reported that the NASA-TLX produced more consistent workload estimates for participants performing the same task than the SWAT technique (Reid and Nygren, 1988) did. Hart and Staveland (1988) also reported that the NASA-TLX workload scores suffer from substantially less between-rater variability than one-dimensional workload ratings did. Pretorious and Cilliers (2007) raised a number of objections to the reliability of the NASA-TLX, arguing that the method provides a measure of participants' subjective opinions of task difficulty and as such is impacted by factors such as experience, personality and even motivation, rather than being an objective measure of task demands.

8.2.12 Tools Needed

A NASA-TLX analysis can be conducted using either pen and paper or the software method. Both the pen-and-paper method and the software method can be purchased from NASA Ames Research Center.

8.2.13 Recommended Text(s)

Hart, S. G., & Staveland, L. E. (1988). Development of a multi-dimensional workload rating scale: Results of empirical and theoretical research. In P. A. Hancock, & N. Meshkati (Eds.), *Human mental workload.* Amsterdam: Elsevier.
Vidulich, M. A., & Tsang, P. S. (1986). Technique of subjective workload assessment: A comparison of SWAT and the NASA bipolar method. *Ergonomics*, 29, 1385–1398.

8.3 The Bedford Scale

8.3.1 Background and Applications

The Bedford scale (Roscoe and Ellis, 1990) is a unidimensional MWL assessment technique that was developed to assess pilot workload. The technique is a very simple one, involving the use of a hierarchical decision tree to assess participant workload via an assessment of spare capacity whilst performing a task. Participants simply follow the decision tree to derive a workload rating for the task under analysis. A scale of 1 (low MWL) to 10 (high MWL) is used. The scale is normally completed posttrial, but it can also be administered during task performance.

8.3.2 Domain of Application

Generic.

8.3.3 Application in Land Use Planning and Urban Design

The Bedford scale may be used to understand the MWL required for any land use planning procedural or analysis task, or indeed to explore the workload associated with a particular urban environment. For example, it is a quick and easy way to understand the user effort required to navigate, or use, a particular urban setting and to understand the different workloads required within an environment by users of different abilities or backgrounds. The range of tasks to be undertaken within that setting may then be explored individually (e.g., crossing the road, accessing information, wayfinding), providing a detailed interpretation of the MWL associated with that environment.

8.3.4 Procedure and Advice

Step 1: Define the Task under Analysis

The first step in a Bedford scale analysis (aside from the process of gaining access to the required systems and personnel) is to define the task (or tasks) that is to be subjected to analysis. The types of task analyzed are dependent upon the focus of the analysis. For example, when assessing the effects on user MWL caused by a novel urban design or a new land use planning process, it is useful to analyze a set of tasks that are as representative of the full functionality of the interface, environment, or procedure as possible. To analyze a full set of tasks will often be too time-consuming and labor-intensive, and so it is pertinent to use a set of tasks that utilize all aspects of the system under analysis.

Step 2: Conduct an HTA for the Task under Analysis

Once the task (or tasks) under analysis is defined clearly, an HTA should be conducted for each task. This allows the analyst and participants to understand the task fully.

Step 3: Selection of Participants

Once the task (or tasks) under analysis is defined, it may be useful to select the participants that are to be involved in the analysis. This may not always be necessary, and it may suffice to simply select participants randomly on the day. However, if workload is being compared across different user types (demographics, physical ability, etc.) then clearly effort is required to select the appropriate participants.

Step 4: Brief Participants

Before the task (or tasks) under analysis is performed, all the participants involved should be briefed regarding the purpose of the study and the Bedford scale technique. It is recommended that participants are given a workshop on MWL and MWL assessment. It may also be useful at this stage to take the participants through an example of Bedford scale analysis so that they understand how the technique works and what is required of them as participants. It may even be pertinent to get them to perform a small task and then to complete a Bedford scale questionnaire. This acts as a *pilot run* of the procedure highlighting any potential problems.

Step 5: Task Performance

Once the participants fully understand the Bedford scale technique and the data collection procedure, they are free to undertake the task under analysis as normal.

Step 6: Completion of the Bedford Scale

Once the participants have completed the relevant task, they should be given the Bedford scale and instructed to work through it, based on the task that they have just completed. Once they have finished working through the scale, a rating of participant MWL is derived. If there are any tasks requiring analysis left, the participants should then move on to the next task and repeat the procedure.

8.3.5 *Advantages*

- The Bedford scale is very quick and easy to use, requiring minimal analyst training.
- It is generic and so can easily be applied in different domains.
- It may be useful when used in conjunction with other techniques of MWL assessment.
- It offers a low level of intrusiveness.

8.3.6 *Disadvantages*

- There is limited validation evidence associated with the technique.
- Its output is limited.
- Participants are not efficient at reporting mental events *after the fact*.

8.3.7 Flowchart

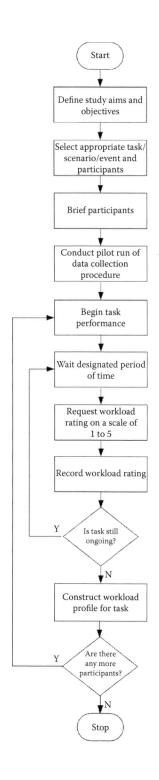

8.3.8 Related Methods

The Bedford scale technique is one of a number of subjective MWL assessment techniques. It is especially similar to the MCH technique (Casali and Wierwille, 1983), as it uses a hierarchical decision tree to derive a measure of participant MWL. When conducting a Bedford scale analysis, a task analysis (such as an HTA) of the task or scenario is normally required.

8.3.9 Approximate Training and Application Times

The training and application times for the Bedford scale are estimated to be very low.

8.3.10 Reliability and Validity

There are no data regarding the reliability and validity of the technique available in the literature.

8.3.11 Tools Needed

The Bedford scale technique is applied using pen and paper.

8.3.12 Example

A modified Bedford scale applicable to the LUP & UD context is presented in Figure 8.4.

8.3.13 Recommended Text(s)

Roscoe, A., & Ellis, G. (1990). *A subjective rating scale for assessing pilot workload in flight*. Farnborough: RAE.

8.4 Instantaneous Self-Assessment

8.4.1 Background and Applications

The ISA workload technique is another very simple subjective MWL assessment technique that was developed by National Air Traffic Services for use in the assessment of air traffic controller MWL during

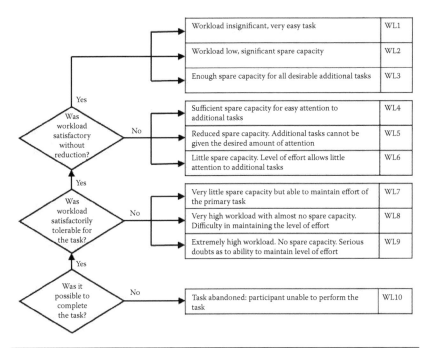

Figure 8.4   Modified Bedford scale of MWL. (Adapted from Roscoe, A. and Ellis, G., *A Subjective Rating Scale for Assessing Pilot Workload in Flight*, RAE, Farnborough, UK, 1990.)

the design of future ATM systems (Kirwan et al., 1997). It involves participants self-rating their workload during a task (normally every two minutes) on a scale of 1 (low) to 5 (high). Kirwan et al. (1997) used the following ISA scale in Table 8.1 to assess air traffic controllers' workload.

Typically, the ISA scale is presented to the participants in the form of a color-coded keypad. The keypad flashes when a workload rating is required, and the participant simply pushes the button that corresponds to their perceived workload rating. It can also be downloaded as digital media application or app. Alternatively, the workload ratings can be requested and acquired verbally. The ISA technique allows a profile of user or participant workload throughout the task to be constructed and allows the analyst to ascertain excessively high or low workload parts of the task under analysis. The appeal of the ISA technique lies in its low resource usage, its low level of intrusiveness, and its flexibility.

Table 8.1   Example ISA Workload Scale

LEVEL	WORKLOAD HEADING	SPARE CAPACITY	DESCRIPTION
5	Excessive	None	Behind on tasks; losing track of the full picture.
4	High	Very little	Nonessential tasks suffering. Could not work at this level very long.
3	Comfortable busy pace	Some	All tasks well in hand. Busy but stimulating pace. Could keep going continuously at this level.
2	Relaxed	Ample	More than enough time for all tasks. Active on ATC task less than 50% of the time.
1	Underutilized	Very much	Nothing to do. Rather boring.

Source: Kirwan, B. et al., *Human Factors in the ATM System Design Life Cycle*, FAA/Eurocontrol ATM R&D Seminar, Paris, France, pp. 16–20, June, 1997.

8.4.2 Domain of Application

Generic.

8.4.3 Application in Land Use Planning and Urban Design

ISA as a workload assessment is inherently useful for designing better and more efficient urban environments. It may be used to understand the workloads of land use planning procedures, processes, and even assessing the effort required in understanding, for example, public documentation or master plans. Here also, it is a technique that lends itself to understanding the workload required in the use of a range of urban environments, by a range of different users. ISA as a method may also be used to capture subjective rating associated with aesthetics, or user experience, or indeed both the cognitive and physical efforts required for wayfinding or walkability.

8.4.4 Procedure and Advice

Step 1: Construct a Task Description

The first step in any workload analysis is to develop a task description for the task or scenario under investigation. It is recommended that an HTA is used for this purpose.

Step 2: Brief Participants

The participants should be briefed regarding the ISA technique, including what it measures and how it works. It may be useful to demonstrate an ISA data collection exercise for a task similar to the one under analysis. This allows the participants to understand how the technique works and also what is required of them. It is also crucial at this stage that the participants have a clear understanding of the ISA workload scale being used. In order for the results to be valid, the participants should have the same understanding of each level of the workload scale, that is, what level of perceived workload constitutes a rating of 5 on the ISA workload scale and what level constitutes a rating of 1. It is recommended that the participants are taken through the scale, and examples of workload scenarios are provided for each level on the scale. Once the participants fully understand the ISA workload scale being used, the analysis can proceed to the next step.

Step 3: Pilot Run

Once the participants have a clear understanding of how the ISA technique works and what is being measured, it is useful to perform a pilot run. Whilst performing a small task, participants should be subjected to the ISA technique. This allows them to experience the technique in a task performance setting. Participants should be encouraged to ask questions during the pilot run to understand the technique and the experimental procedure fully.

Step 4: Begin Task Performance

Next, the participants should begin the task under analysis. A simulation of the environment under may be used; however, this is dependent upon the domain of application.

Step 5: Request and Record Workload Rating

The analyst may request a workload rating either verbally, through flashing lights on the workload scale display, or through the app if the data are being collected in situ. The frequency and timing of the workload ratings should be determined beforehand by the analyst. Typically, a workload rating is requested every 2 min. It is crucial that the provision of a workload rating is as unintrusive to the participant's primary task performance as possible.

Step 4 should continue at regular intervals until the task is completed. The analyst should make a record of each work-load rating given.

Step 6: Construct Task Workload Profile

Once the task is complete and the workload ratings are collected, the analyst should construct a workload profile for the task under analysis. Typically, a graph is constructed, highlighting the high and low workload points of the task under analysis. An average workload rating for the task under analysis can also be calculated.

8.4.5 *Advantages*

- ISA is a very simple technique to learn and use.
- The output allows a workload profile for the task under analysis to be constructed.
- It is very quick in its application as data collection occurs during the trial.
- It has been used extensively in numerous domains.
- It requires very little in the way of resources.
- Although it is obtrusive to the primary task, it is probably the least intrusive of the online workload assessment techniques.
- It is a low-cost technique.

8.4.6 *Disadvantages*

- ISA is intrusive to primary task performance.
- There is limited validation evidence associated with the technique.

- It is a very simplistic technique, offering only a limited assessment of workload.
- Participants are not very efficient at reporting mental events.

8.4.7 Related Methods

ISA is a subjective workload assessment technique of which there are many scales, such as the NASA-TLX, and Bedford. To ensure comprehensiveness, it is often used in conjunction with other subjective techniques, such as the NASA-TLX.

8.4.8 Approximate Training and Application Times

It is estimated that the training and application times associated with the ISA technique are very low. Application time is dependent upon the duration of the task under analysis.

8.4.9 Reliability and Validity

Djokic et al. (2010) used ISA to explore the ATC operators' subjective workload in a simulated experiment involving the Central European Air Traffic Services airspace. The subjective measure of workload was compared with ATC complexity components (identified using principal component analysis) and controller activity metrics to assess the relative contribution of each to workload. The results found that complexity components alone could not predict workload as additional factors contributed to workload level such as communication load.

8.4.10 Tools Needed

ISA can be applied using pen and paper.

8.4.11 Flowchart

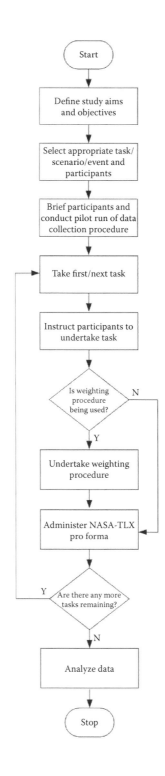

8.4.12 Recommended Text(s)

Kirwan, B., Evans, A., Donohoe, L., Kilner, A., Lamoureux, T., Atkinson, T., & MacKendrick, H. (1997). *Human factors in the ATM system design life cycle.* Paris: FAA/Eurocontrol ATM R&D Seminar, June 16–20, 1997.

9

INTERFACE EVALUATION METHODS

9.1 Introduction

Urban and regional environments represent an array of unique natural and human-made interfaces. The built environment represents an interface established by land use planning and urban design (LUP & UD) policy and practitioners, which mediates between the needs of humans in the space and the expectations and boundaries of urban planning. When we take the perspective of our urban settings as interfaces, it is then possible to apply a range of human factors (HF) methodologies and gain greater insights into their usability, satisfaction, and overall design. The output of interface analysis methods is typically used to improve the interface in question through redesign. For example, when considering human–computer interactions, ISO 9241-11 requires that the usability of software is considered along three dimensions: effectiveness (how well does the product performance meet the tasks for which it was designed?), efficiency (how much time or effort is required to use the product to perform these tasks?), and attitude (e.g., how favorably do users respond to the product?). Such enquires and the more rigorous exploration of the array of human interfaces within the LUP & UD disciplines is certainly warranted.

Further there may be practical lessons to be learned from HF when considering ISO 9241-210:2010 (human-centered design for interactive systems). This standard details the need to apply interface-analysis techniques throughout a product's life cycle, either in the design stage to evaluate design concepts or in the operational stage to evaluate effects on performance. In particular, this standard calls for the active involvement of users in the design process to gain an appropriate understanding of requirements and an appropriate allocation of function between users and the products. It assumes that the design

253

process is both multidisciplinary and iterative. This suggests that there is a need to have a clear and consistent set of representations that can be shared across the design team and revised during the development of the design. In the current chapter, we review methods that can fulfill these requirements. Most of the methods considered in this review require at least some form of interface, ranging from conceptual designs and diagrams to the operational environment itself, and most methods normally use end users of the system under analysis.

A number of different types of interface-analysis technique are available, such as usability assessment, error analysis, interface layout analysis, and general interface assessment techniques. Usability-assessment techniques are used to assess the usability (effectiveness, learnability, flexibility, and attitude) of a particular interface. Typically, these are completed by potential end users based upon user trials with the environment or system under analysis.

The layout of an interface can also be assessed using techniques such as link analysis and layout analysis. As the names suggest, these methods are used to assess the layout of the interface and its effects upon task performance. More general interface-analysis techniques such as heuristic evaluation (including those of Neilsen, 1994; Schneiderman, 1998) and user trials are used to assess the interface as a whole and are flexible in that the focus of the analysis is determined by the analyst. The advantages associated with the use of interface-analysis techniques lie in their simplistic nature and the usefulness of their outputs. Most of the techniques are simple to apply, requiring minimal time and costs, and also require only minimal training. The utility of the outputs is also ensured, as most approaches offer interface redesigns based upon end-user opinions. The only significant disadvantages associated with the use of interface-analysis techniques are that the data-analysis procedures may be time-consuming and laborious, and also that much of the data obtained is subjective. A brief description of the interface-analysis methods considered in this review is given in the following.

Checklists offer a simplistic and low-cost approach to interface assessment. When using a checklist, the analyst checks the environment or system interface against a predefined set of criteria to evaluate its usability. *Heuristics* analysis is one of the simplest interface-analysis techniques available, involving simply obtaining subjective opinions of the analyst based on his or her interactions with a particular environment or design. In conducting a heuristic analysis, an analyst or end user should perform a user trial with the environment, design, or urban artifacts (seating, signage, shelter, etc.) under analysis and make observations regarding the usability, quality, and error potential of the design.

Link analysis is used to evaluate and redesign an interface in terms of the nature, frequency, and importance of links between elements of the interface in question. A link analysis defines links (physical movements, hand or eye movements) between elements of the interface under analysis. The interface is then redesigned on the basis of these links, with the most often linked elements of the interface relocated to increase their proximity to one another. *Layout analysis* is also used to evaluate and redesign the layout of the interface in question. This involves arranging the interface components into functional groupings and then organizing these groups by importance of use, sequence of use, and frequency of use. The layout-analysis output offers can offer an urban redesign based upon a variety of user's interactions with a task.

9.2 Checklists

9.2.1 Background and Applications

Checklists offer a quick, easy, and low-cost approach to interface assessment. Typical checklist approaches involve analysts checking features associated with a product or interface against a checklist containing a predefined set of criteria. Checklist style evaluation can occur throughout the life cycle of a product or system, from CAD or paper drawings to the finished product. Checklists can be used to evaluate the usability and design of a system in any domain. In the past, they have been used to evaluate product usability in the human–computer interaction (HCI) (Ravden and Johnson, 1989), automotive (Stanton and Young, 1999a), and ATC domains. When using checklists, the analyst should have some level of skill or familiarity with the environment or artifact under evaluation. Performing a checklist analysis is a matter of simply inspecting the artifact against each point on an appropriate checklist. Checklists are also very flexible in that they can be adapted or modified by the analyst according to the demands of the investigation.

9.2.2 Domain of Application

Generic. Although checklist techniques are originated in the HCI domain, they are typically generic and can be applied in any domain.

9.2.3 Application in Land Use Planning and Urban Design

LUP & UD researchers and practitioners are familiar with checklists, yet often disregard them as a relevant method of evaluation. They in fact offer significant insights into the appropriate design and redesign of both the elements within an environment as well as the environment itself for a range of users and under different temporal and microclimate conditions.

9.2.4 Procedure and Advice

Step 1: Select the appropriate checklist

First, the analyst must decide which form of checklist is appropriate for the artifact, environment, or system under investigation. The checklist used may be simply an existing one or the analyst may choose to adapt an existing checklist to make it more appropriate for the system under analysis. Alternatively, if a suitable checklist is not available, the analyst may choose to create a new checklist specifically for the system/product in question.

Step 2: Check item on checklist against product

The analyst should take the first point on the checklist and check it against the artifact, environment, or system under analysis. For example, the first item on a planning checklist for cycling pathways: "Are the paths a minimum of 2.4 m wide?" The analysts should then proceed to check the width of the pathway. The options given may be *Always*, *Most of the time*, *Some of the time*, and *Never*. By using subjective judgment, the analyst should rate the environment under analysis according to the checklist item. Step 2 should be repeated until each item on the checklist has been dealt with.

9.2.5 *Flowchart*

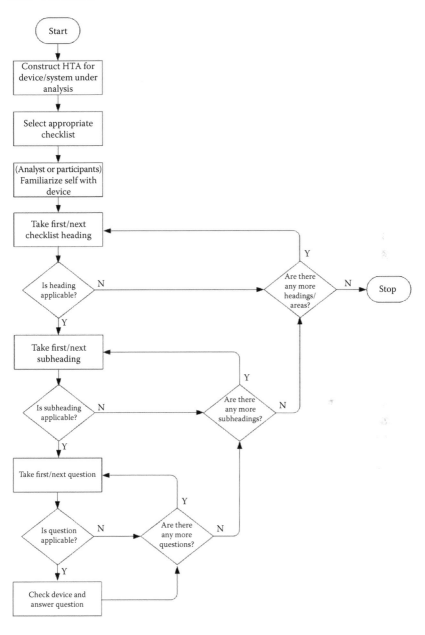

9.2.6 Advantages

- Checklists are quick and simple to apply and incur only a minimal cost.
- They offer an immediately useful output.
- They are based upon established knowledge about human performance (Stanton and Young, 1999a).
- The technique requires very little training.
- Resource usage is very low.
- They are very adaptable and can easily be modified to use for many scenarios.
- A number of different checklists are available to the LUP & UD practitioner.

9.2.7 Disadvantages

- A checklist-type analysis does not account for errors or cognitive problems associated with the device.
- Context is ignored by checklists.
- Checklist data are totally subjective. What one analyst classes as bad design may be classed as suitable by another.
- Checklists offer a low level of consistency.
- They are not a very sophisticated approach to system evaluation.

9.2.8 Example

The following example (Figure 9.1) is an extract of a Sidewalks Implementation Strategy Checklist (City of Los Angeles 2008, p. 9) that is supported by illustrative recommendations of practice.

Sidewalks implementation strategy checklist

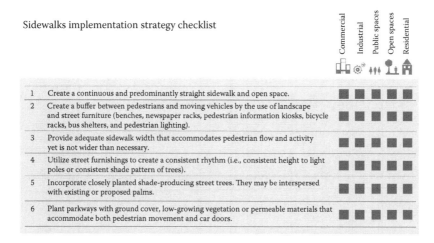

	Commercial	Industrial	Public spaces	Open spaces	Residential
1	Create a continuous and predominantly straight sidewalk and open space.	■	■	■	■	■
2	Create a buffer between pedestrians and moving vehicles by the use of landscape and street furniture (benches, newspaper racks, pedestrian information kiosks, bicycle racks, bus shelters, and pedestrian lighting).	■	■	■	■	■
3	Provide adequate sidewalk width that accommodates pedestrian flow and activity yet is not wider than necessary.	■	■	■	■	■
4	Utilize street furnishings to create a consistent rhythm (i.e., consistent height to light poles or consistent shade pattern of trees).	■	■	■	■	■
5	Incorporate closely planted shade-producing street trees. They may be interspersed with existing or proposed palms.	■	■	■	■	■
6	Plant parkways with ground cover, low-growing vegetation or permeable materials that accommodate both pedestrian movement and car doors.	■	■	■	■	■

1 Create a continuous and predominantly straight sidewalk and open space.

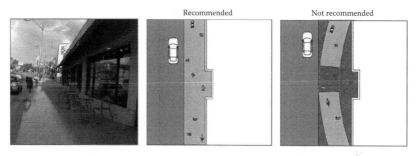

Recommended   Not recommended

Figure 9.1 Sidewalks implementation strategy checklist extract.

9.2.9 Related Methods

There are a number of checklists available to the LUP & UD and the HF practitioner.

9.2.10 Approximate Training and Application Times

Checklists require only minimal training time. Similarly, the application time associated with checklist techniques is minimal. In an analysis of 12 ergonomics methods, Stanton and Young (1999a) reported that checklists are one of the quickest techniques to train, practice, and apply.

9.2.11 Reliability and Validity

Although Stanton and Young (1999a) reported that checklists performed quite poorly on intrarater reliability, they also report that inter-rater reliability and predictive validity of checklists was good.

9.2.12 Tools Needed

Checklists can be applied using pen and paper only; however, for a checklist analysis, the analyst must have access to some example of a design or urban environment under analysis. An appropriate checklist is also required.

9.2.13 Recommended Text(s)

Ravden, S. J., & Johnson, G. I. (1989). *Evaluating usability of human-computer interfaces: A practical method*. West Sussex: Ellis Horwood.
Stanton, N. A., & Young, M. S. (1999). *A guide to methodology in ergonomics: Designing for human use*. London: Taylor & Francis Group.

9.3 Heuristic Analysis

9.3.1 Background and Applications

Heuristic-analysis techniques offer a quick and simple approach to interface evaluation. Heuristic analysis involves analysts providing subjective opinions based upon their interaction with a particular design, urban interface, or environment. It is a flexible approach that can be used to assess a number of features associated with a particular product or interface, including usability, error potential, mental workload (MWL), and overall design quality. To conduct a heuristic analysis, an analyst (or team of analysts) simply performs a series of interactions with the design, interface, or environment under analysis, recording their observations as they proceed. Heuristic-type analyses are typically conducted throughout the design process to evaluate design concepts and propose remedial measures for any problems encountered.

The popularity of heuristic analysis lies in its simplicity and the fact that it can be conducted easily and with only minimal resource usage at any stage throughout the design process.

9.3.2 Domain of Application

Generic.

9.3.3 Application in Land Use Planning and Urban Design

Heuristic analysis can be used within an LUP & UD context to evaluate any technology, devices, and performance aids that may be deployed within urban settings or used procedurally. It can be used throughout the design process to evaluate design concepts or for the evaluation of operational urban environments.

9.3.4 Procedure and Advice

Step 1: Define the task under analysis
> The first step in a heuristic analysis is to define a representative set of tasks or scenarios for the design, interface, or environment under analysis. It is recommended that heuristic analyses are based upon the analyst performing an exhaustive set of tasks with the design in question. The tasks defined should then be placed in a task list. It is normally useful to conduct a hierarchical task analysis (HTA) for this purpose, based on the operation of the device in question. The HTA then acts as a task list for the heuristic analysis.

Step 2: Define heuristic list
> In some cases, it may be fruitful to determine which aspects are to be evaluated before the analysis begins. Typically, usability (ease of use, effectiveness, efficiency, and comfort) and error potential are evaluated.

Step 3: Familiarization phase
> To ensure that the analysis is as comprehensive as possible, it is recommended that the analyst involved spends some time familiarizing themselves with the design in question. This might involve consultation with the associated design

documentation or being taken through a walkthrough of the concept in question.

Step 4: Perform tasks

Once familiar with the environment under analysis, the analyst should then perform each task from the task list developed during Steps 1 and 2 and offer opinions regarding the design and the heuristic categories required. During this stage, any good points or bad points associated with the participant's interactions with the design should be recorded. If the analysis concerns a design concept, then a task walkthrough is sufficient. Each opinion offered should be recorded.

Step 5: Propose remedies

Once the analyst has completed all of the tasks from the task list, remedial measures for any of the problems recorded should be proposed and recorded.

9.3.5 *Advantages*

- Heuristic analysis offers a quick, simple, and low-cost approach to usability assessment.
- Due to its simplicity, only minimal training is required.
- It can be applied to any form of product, including paper-based diagrams, mock-ups, prototype designs, and functional devices.
- The output derived is immediately useful, highlighting problems associated with the design in question.
- It involves the use of very few resources.
- It can be used repeatedly throughout the design life cycle.

9.3.6 *Disadvantages*

- Heuristic analysis offers poor reliability, validity, and comprehensiveness.
- It requires subject matter experts in order for the analysis to be worthwhile.
- It is subjective.
- It is totally unstructured.
- The consistency of such a technique is questionable.
- It is difficult to perform (Bennett and Stephens, 2009).

9.3.7 Example

There are many forms of heuristic analysis, Cronholm (2009) argues that both Schneiderman's eight golden rules (Schneiderman, 1998) and Nieslen's 10 heuristics (Nielsen, 1994a, b) are frequently cited within the literature, illustrating a large impact on interface development. The importance of these two sets of heuristics is continually emphasized by researchers (Singh and Wesson, 2009). From these sets of HCI heuristics, the authors here have established an example set of LUP & UD appropriate heuristics, Table 9.1, which will provide guidance in the analysis of urban systems and designs.

Table 9.1 Example of Heuristics-Appropriate Land Use Planning and Urban Design Interface Design and Evaluation

Offer feedback/visibility of system status	The system should always keep users informed through appropriate feedback within a reasonable time. The level of feedback should match the level of task (i.e., simple/small task = modest/quick feedback) (Neilsen 1994a, b; Schneiderman 1998)
Match between system and the real world	The system should speak the user's language, with words, phrases, and concepts familiar to the user, rather than system-oriented terms. It should follow real-world conventions, making information appear in a natural and logical order (Neilsen 1994a, b)
User control and freedom	Offer simple error handling and reverse actions. Users often choose system functions by mistake and will need a clearly marked *emergency exit*. Support undo and redo (Neilsen 1994a, b; Schneiderman 1998)
Consistency and standards	Users should not have to wonder whether different words, situations, or actions mean the same thing. Follow platform conventions (Neilsen 1994a, b)
Recognition rather than recall	Minimize the user's memory load by making objects, actions, and options visible. The user should not have to remember information from one action to another. Make users initiators, not responders of action (Neilsen 1994a, b; Schneiderman 1998)
Flexibility and efficiency of use. Develop shortcuts	Accelerators—unseen by the novice user—may often speed up the interaction for the expert user such that the system can cater to both inexperienced and experienced users. Users should be permitted to tailor frequent actions (Neilsen 1994a, b; Schneiderman 1998)
Aesthetic and minimalist design	Designs and dialogs should not contain information that is irrelevant or rarely needed. Keep it beautiful and uncomplicated (Neilsen 1994a, b; Schneiderman 1998)
Help and instruction	Even though it is better if the system can be used without documentation, it may be necessary to provide help and instruction (Neilsen 1994a, b)

9.3.8 Approximate Training and Application Times

The technique requires very little, if any, training, and the associated application time is also typically low. Comparing a number of techniques in the analysis of in-car interfaces, Harvey and Stanton (2013) concluded that heuristic analysis required fewer time resources than methods such as HTA, systematic human-error reduction and prediction approach (SHERPA), and critical path analysis. They estimated that the method required 1 h for data collection and 1 h for analysis.

9.3.9 Reliability and Validity

In conclusion to their comparison of 12 HF methods, Stanton and Young (1999b) reported that the unstructured nature of the technique led to very poor results for reliability and predictive validity. Both intra- and interanalyst reliability for the technique are questionable, due to its unstructured nature.

9.3.10 Tools Needed

Heuristic analysis is conducted using pen and paper only. The environment or design under analysis is required in some form, for example, functional diagrams, the actual design, or paper drawings.

9.3.11 Recommended Text(s)

Stanton, N. A., & Young, M. S. (1999). *A guide to methodology in ergonomics: Designing for human use*. London: Taylor & Francis Group.

9.4 Link Analysis

9.4.1 Background and Applications

Link analysis is an interface evaluation method that is used to identify and represent *links* in a system between interface components and operations and to determine the nature, frequency, and importance of these links. In HF, links are defined as movements of attentional gaze or position between parts of the system, or communication with other system elements. For example, if an actor is required to press button A and then button B in sequence to accomplish a particular task, a link between buttons A and B is recorded. Link analysis uses spatial dia-grams to represent the links within the system or device under analy-sis, with each link represented by a straight line between the *linked* interface elements. Specifically aimed at aiding the design of inter-faces and systems, the most obvious use of link analysis is in the area of workspace-layout optimization (Stanton and Young, 1999a), that is, the placement of controls and displays according to their importance first, then to their frequency of use, then to their function within the system, and finally to their sequence of use. Link analysis was origi-nally developed for use in the design and evaluation of process control rooms (Stanton and Young, 1999a), but it can be applied to any system in which the user exhibits hand or eye movements. When conducting a link analysis, establishing the links between system/interface com-ponents is normally achieved through a walkthrough or observational study of the task (or tasks) under analysis. The output of a link analy-sis is normally a link diagram and also a link table (both depict the same information). The link diagram and table can be used to suggest revised layouts based on the premise that links should be minimized in length, particularly if they are important or frequently used.

9.4.2 Domain of Application

Generic.

9.4.3 Application in Land Use Planning and Urban Design

Link analysis may allow the design of better interfaces in a range of LUP & UD scenarios. The interaction with and the comprehension of our built environments through wayfinding are central to their use and usability. When thinking about urban design, it offers a way in which to design appropriate use of space at important decision-making *nodes* (Lynch, 1960) within a city systems (e.g., public transport stops; street crossings; changes in levels, bike hubs). Link analysis can reveal what users require and look for at these junc-tures (information, comfort, amenity, shelter), how do they need it, and exactly when. There are new possibilities for urban design when efficient and intuitive systems within systems are designed and deployed.

9.4.4 Procedure and Advice

Step 1: Define the task under analysis

The first step in a link analysis involves clearly defining the task (or tasks) under analysis. When using link analysis to evaluate the interface layout of a particular device or system, it is recommended that a set of tasks that are as representative of the full functionality of the device or system are used. It is normally useful to conduct an HTA for normal operation of the device or system in question at this stage, as the output can be used to specify the tasks that are to be analyzed. It also allows the analyst involved to gain a deeper understanding of the tasks and the device under investigation.

Step 2: Task analysis/list

Once the task (or tasks) under analysis is clearly defined, a task list including (in order) all of the component task steps involved should be created. The task list can be derived from the HTA. Typically, a link analysis is based upon the bottom-level tasks or operations identified in the HTA developed during Step 1.

Step 3: Data collection

The analyst should then proceed to collect data regarding the tasks under analysis. This normally includes performing a walkthrough of the task steps contained in the task list and also observational study of the task in question. The analyst should record which components are linked by hand/eye movements and how many times these links occur during the tasks performed.

Step 4: Construct link diagram

Once the data collection phase is complete, construction of the link diagram can begin. This involves creating a schematic layout of the device/system/interface under analysis and add-ing the links between interface elements recorded during the data collection phase. Links are typically represented in the form of lines joining the linked interface elements or compo-nents. The frequency of the links is represented by the num-ber of lines linking each interface element, for example, seven lines linking interface elements A and B represent a total of seven links between the two interface elements during the task under analysis.

Step 5: Link table

The link diagram is accompanied by a link table, which displays the same information as the link diagram, only in a tabular format. Components take positions at the heads of the rows and columns, and the numbers of links are entered in the appropriate cells.

Step 6: Redesign proposals

Although not compulsory as part of a link analysis, a redesign for the interface under analysis is normally offered on the basis of the links defined between the interface elements during the analysis. The redesign is designed to reduce the distance between the linked interface components, particularly the most important and frequently used linked components.

9.4.5 *Advantages*

- Link analysis is a very simple technique that requires only minimal training.
- It is a quick technique that offers an immediately useful output.
- Its output helps one to generate design improvements.
- It has been used extensively in the past in a number of domains.
- Its output prompts logical redesign of system interfaces.
- It can be used throughout the design process to evaluate and modify design concepts.

9.4.6 *Disadvantages*

- Link analysis requires preliminary data collection, including observational study and a walkthrough analysis of the task under analysis.
- The development of an HTA adds considerable time to the analysis.
- It only considers the basic physical relationship between the user and the system. Cognitive processes and error mechanisms are not accounted for.
- Its output is not easily quantifiable.

9.4.7 Flowchart

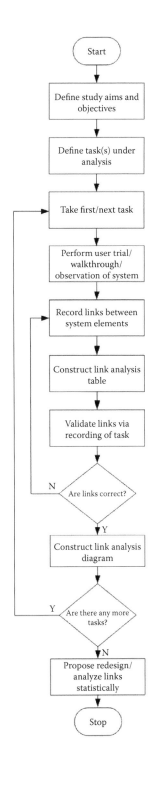

9.4.8 Example

The following example presents the results of a link analysis of the interfaces associated with a pedestrian activated midblock pedestrian crossing. Table 9.2 identifies the tasks and crossing components associated with a pedestrian approaching and waiting at the crossing, in addition to them entering and exiting the crossing. Also established is the type of link—visual or physical—whereas the components in Table 9.3 are identified for each part of the system; and their links during the task are shown in Table 9.4. Figure 9.2 is the link diagram for crossing the road, identifying both the number of links and if they were visual or physical links.

In this example, the links have been established only from the perspective of the pedestrian. However, in a complex system such as this, it is possible to identify a number of other user and component interfaces, including cyclists and motorists. This type of multiuser analysis of systems is critical to the success and efficiency of urban environments.

Table 9.2 Pedestrian Crossing Tasks Associated with Crossing Components

APPROACH AND WAIT AT THE CROSSING			
HTA TASKS	CROSSING COMPONENTS	VISUAL LINK	PHYSICAL LINK
Check traffic right	Traffic right	X	
Check traffic left	Traffic left	X	
Check traffic signal	Traffic signal	X	
Check walk signal	Walk signal	X	
Press signal button	Signal button		X
Check sidewalk waiting point	Sidewalk waiting point	X	
Check position of other pedestrians	Sidewalk waiting point	X	
Check crossing surface	Crossing surface	X	
Recheck walk signal	Walk signal	X	
ENTER AND LEAVE THE CROSSING			
HTA TASKS	INTERFACE ELEMENT	VISUAL LINK	PHYSICAL LINK
Check walk signal	Walk signal	X	
Check traffic right	Traffic right	X	
Check traffic left	Traffic left	X	
Check position of other pedestrians	Sidewalk waiting points	X	
Check crossing surface	Crossing surface	X	
Enter crossing	Crossing surface		X
Leave crossing	Sidewalk waiting points		X

Table 9.3   Component Table of the System

A. Traffic right	E. Signal button
B. Traffic left	F. Sidewalk waiting point
C. Traffic signal	G. Sidewalk waiting point
D. Walk signal	H. Crossing surface

Table 9.4   Link Table for Crossing the Road

	A	B	C	D	E	F	G	H
A	x							
B		x						
C			x					
D				x				
E					x			
F	2	2	1	2	1	x	1	2
G							x	1
H								x

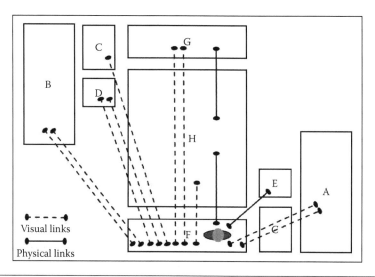

Figure 9.2   Link diagram for pedestrian interface with midblock roadway crossing.

9.4.9 Related Methods

A link analysis normally requires an initial task description to be created for the task under analysis, such as an HTA. In addition, an observation or walkthrough analysis of the task under analysis should be performed to establish the links between components in the system.

9.4.10 Approximate Training and Application Times

In conclusion to their comparison of 12 ergonomics methods, Stanton and Young (1999a) reported that the link-analysis technique is relatively fast to train and practice, and also that execution time is moderate compared with the other techniques (e.g., SHERPA, layout analysis, repertory grids, and checklists).

9.4.11 Reliability and Validity

Stanton and Young (1999a) reported that link analysis performed particularly well on measures of intrarater reliability and predictive validity. They also reported, however, that the technique was let down by poor inter-rater reliability.

9.4.12 Tools Needed

When conducting a link analysis, the analyst should have the device under analysis, pen and paper, and a stopwatch. For the observation part of the analysis, a video-recording device is required. An eye-tracker device can also be used to record fixations during the task performance.

9.4.13 Recommended Text(s)

Drury, C. G. (1990). Methods for direct observation of performance. In Wilson, J. and Corlett, E. N. (Eds.). *Evaluation of human work: A practical ergonomics methodology*, 2nd edition, pp. 45–68. London: Taylor & Francis Group.

Stanton, N. A, & Young, M. S. (1999). *A guide to methodology in ergonomics: Designing for human use.* London: Taylor & Francis Group.

9.5 Layout Analysis

9.5.1 Background and Applications

Layout analysis is similar to link analysis and in that it is based on spatial diagrams of the product and its output directly addresses interface design. It is used to analyze existing designs and suggests improvements to the interface arrangements based on functional grouping. The theory behind layout analysis is that the interface should mirror the user's structure of the task and the conception of the interface as a task map greatly facilitates design (Easterby, 1984). A layout analysis begins by simply arranging all of the components of the interface into functional groupings. These groups are then organized by their importance of use, sequence of use, and frequency of use. The components within each functional group are then reorganized; once again, this is done according to importance, sequence, and frequency of use. The components within a functional group will then stay in that group throughout the analysis and cannot move anywhere else in the reorganization stage. At the end of the process, the analyst has redesigned the device in accordance with the user's model of the task based upon importance, sequence, and frequency of use.

9.5.2 Domain of Application

Generic.

9.5.3 Application in Land Use Planning and Urban Design

To date, LUP & UD practitioners have undertaken the kinds of processes identified here as layout analysis more intuitively. That is the organization of urban spaces into functional groupings to allow for greater efficiencies in the design and use of our cities. Here, however, is a detailed description and theoretical background to these processes. These HF insights afford early career and established LUP & UD researchers and practitioners a grounding and replicability in what is an important design activity in our discipline.

9.5.4 Procedure and Advice

Step 1: Schematic diagram

First, the analyst should create a schematic diagram for the environment under analysis. This diagram should contain each (clearly labeled) interface and urban element.

Step 2: Arrange interface components into functional groupings

The analyst begins by arranging the interface components into functional groupings. Each urban element should be grouped according to its function in relation to the environment under analysis. For example, the interface components of a sidewalk were arranged into the functional groups infrastructure, furniture, throughway, and frontage (Stevens and Salmon, 2014).

Step 3: Arrange functional groupings into importance of use

Next, the analyst should arrange the functional groupings into importance of use. The analyst may want to make the most important functional group the most readily available on the interface. Again, this is based on the intended purpose of the system and the analyst's judgment.

Step 4: Arrange functional groupings into sequence of use

The analyst should then repeat Step 3, only this time arranging the functional groupings based on their sequence of use.

Step 5: Arrange functional groupings into frequency of use

The analyst should then repeat Step 3, only this time arranging the functional groupings based on their frequency of use. At the end of the process, the analyst has redesigned the environment according to the end-user map of the task (Stanton and Young, 1999a) and the intended functions of the environment.

Step 6: Redesign the interface

Once the functional groups have been organized on the basis of their importance, sequence, and frequency of use, the process is repeated for within functional group items (organization of elements within the furniture group). The analyst should base the interface redesign on the three categories (importance, sequence, and frequency of use). For example, the analyst may wish to make the most important and frequently used aspect of the interface (pavement) the most readily available.

9.5.5 Flowchart

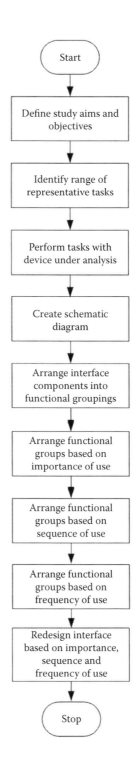

9.5.6 *Advantages*

- Layout analysis offers a quick, easy to use, and low-cost approach to interface design and evaluation.
- It can be used at a range of LUP & UD scales.
- It has a low level of resource usage.
- It requires only minimal training.
- It can be applied to concept maps, schematics, or master plan documents of the environment/interface under analysis.
- Its output provided is immediately useful, offering a redesign of the interface under analysis based on importance, sequence, and frequency of use of the interface elements.

9.5.7 *Disadvantages*

- Layout analysis offers poor reliability and validity (Stanton and Young, 1999a).
- The output of the technique is very limited, that is, it only caters for layout. Errors, MWL, and task performance times are ignored.
- Literature regarding layout analysis is extremely sparse.
- If an initial HTA is required, application time can rise dramatically.
- Conducting a layout analysis for complex interfaces may be very difficult and time-consuming.

9.5.8 *Example*

Here layout analysis has been used to redesign a typical roadside pedestrian and cycleway environment. Stevens et al. (2017) established a Work Domain Analysis that prioritized walking and cycling as the principal modes of urban transport. The outcomes of that analyses are used here to develop the design and use requirements.

Figure 9.3 outlines the development of the street-side environment by way of the layout analysis process. The initial design is first presented as a schematic which is then organized into the functional groupings of task or design elements. Next the importance of use is established. Here the uses are prioritized for pedestrian and cyclist mobility and as such the roadway and car parking functions are less influential. The sequence of use is then detailed, followed by consideration of the frequency of use,

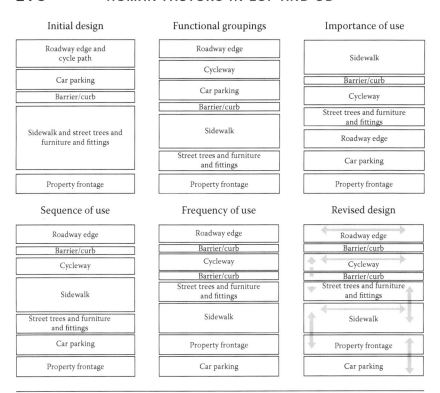

Figure 9.3 Layout analysis of a roadside environment for walking and cycling. (Adapted from Stevens, N. J. et al., Work domain analysis applications in urban planning: Active transport infrastructure and urban corridors. In *Cognitive Work Analysis: Applications, Extensions and the Future*, Taylor & Francis Group, Boca Raton, FL, 2017.)

ultimately resulting in a revised design. The outcome is intended as a pedestrian and cyclist friendly environment, which affords users safety, amenity and flexibility. Incompatible modes of transport are separated allowing the efficiency of movement through the corridor. However, users can also move safely between active modes of transport across the corridor, as boundaries are permeable, yet clearly delineated optimizing the use of shared facilities and street-side amenity.

9.5.9 Related Methods

Layout analysis is very similar to link analysis in its approach to interface design.

9.5.10 Approximate Training and Application Times

In conclusion to their comparison study of 12 ergonomics methods, Stanton and Young (1999a) reported that little training is required for layout analysis, and that it is amongst the quickest of the 12 techniques to apply. It is therefore estimated that the training and application times associated with the technique are low. However, if an initial HTA is required, the application time would rise considerably. Harvey and Stanton (2013) concluded that layout analysis was quicker to use than methods such as HTA, SHERPA, and critical path analysis. They estimated that the method required 1–2 h for data collection and an hour for analysis.

9.5.11 Reliability and Validity

Stanton and Young (1999a) reported poor statistics for intrarater reliability and predictive validity for layout analysis.

9.5.12 Tools Needed

Layout analysis can be conducted using pen and paper, providing the device or pictures of the device under analysis are available.

9.5.13 Recommended Text(s)

Stanton, N. A., & Young, M. S (1999). *A guide to methodology in ergonomics.* London: Taylor & Francis Group.

9.6 Walkthrough Analysis

9.6.1 Background and Applications

Walkthrough analysis is a very simple procedure used by designers whereby experienced users of a system or environment perform a walkthrough or demonstration of a task or set of tasks using the system under analysis. Walkthroughs are typically used early in the design process to envisage how a design concept would work and also to evaluate and modify the design concept. They can also be used on existing urban systems to demonstrate to designers how a process or task is currently performed, highlighting flaws, error potential, and usability problems. The appeal of walkthrough-type analysis lies in the fact that the scenario or task under analysis does not necessarily have to occur. That is, one of the problems of observational study is that the required scenario simply may not occur or, if it does, the observation team may have to spend considerable time waiting for it to occur. Walkthrough analysis allows the scenario to be *acted out*, removing the problems of gaining access to systems and personnel and also waiting for the scenario to occur. A walkthrough involves a participant walking through a scenario, performing (or pretending to perform) and explaining the actions that would occur. The walkthrough is also verbalized and the analyst can stop the scenario and ask questions at any point. Walkthrough analysis is particularly useful in the initial stages of task-analysis development.

9.6.2 Domain of Application

Generic.

9.6.3 Application in Land Use Planning and Urban Design

Walkthrough analyses may allow for the better understanding of a range of land use planning and urban design concepts and urban settings. Many LUP & UD practitioners already incorporate similar methodologies for site analysis and concept development. For example, undertaking a walkthrough analysis of the anticipated tasks and activities on site allows for a better understanding and inclusion of subjective elements and sensory receptors such as view sheds and aural amenity for urban development scenarios.

9.6.4 Procedure and Advice

There are no set rules for a walkthrough analysis. The following procedure is intended to act as a set of guidelines for conducting such an analysis of a proposed (system) design concept.

Step 1: Define the set of representative scenarios

First, a representative set of tasks or scenarios for the system under analysis should be defined. As a general rule, the set of scenarios used should cover every aspect of the design and its interface at least once. The personnel involved in each scenario should also be defined. If the required personnel cannot be gathered for the walkthrough, then members of the design team can be used.

Step 2: Conduct an HTA for scenario under analysis

Once a representative set of tasks for the system or device under analysis are defined, they should be described using an HTA. This involves breaking down the task under analysis into a hierarchy of goals, operations, and plans. Tasks are broken down into hierarchical set of tasks, subtasks, and plans. The HTA is useful as it gives the analyst a clear description of how the task (or tasks) should be carried out and also defines the component task steps involved in the scenario (or scenarios) under analysis.

Step 3: Perform walkthrough

Next, the analyst simply takes each scenario and performs a verbalized walkthrough using the concept design under analysis. It is recommended that the analyst uses the HTA to determine the component task steps involved. The scenario can be frozen at any point and questions can be asked regarding design, comfort, decisions made, situation awareness, potential for errors, and so on. The walkthrough should be recorded using video-recording equipment. Any problems with the design concept encountered during the walkthrough should be recorded and design remedies offered and tested.

Step 4: Analyze data

Once the walkthrough has been performed, the data should be analyzed accordingly and used with respect to the goals of the analysis. Walkthrough data are very flexible and can be used for a number of purposes, such as task analysis, constructing timelines, and evaluating error potential.

Step 5: Modify design

Once the walkthrough is complete and the data are analyzed, the design can be modified on the basis of the remedial measures proposed as a result of the walkthrough. If a new design is proposed, a further walkthrough should be conducted to analyze the new design.

9.6.5 Advantages

- When used correctly, a walkthrough can provide a very accurate description of the task under analysis and also how a proposed system design would be used.

- Walkthrough analysis allows the analyst to stop or interrupt the scenario to query certain points. This is a provision that is not available when using other techniques, such as observational analysis.
- It does not necessarily require the system under analysis and so can be performed in early phases of the design process (Mahatody et al., 2010).
- It is a simple, quick, and low-cost technique.
- It would appear to be a very useful tool in the analysis of distributed (team-based) tasks.
- It can provide a very powerful assessment of a design concept.
- It requires little training (Mahatody et al., 2010).
- It offers a simple, guided process.
- It can identify errors or error potential and redundant features of a system.

9.6.6 Disadvantages

- For the analysis to be fruitful, experienced operators for the system under analysis are required.
- The reliability of the technique is questionable.
- Further guidance on preparation and analysis is required (Mahatody et al., 2010).

9.6.7 Related Methods

The walkthrough technique is very similar to verbal protocol analysis and observational analysis. Cognitive walkthrough is commonly used with think-aloud methods (Peute and Jaspers, 2007).

9.6.8 Approximate Training and Application Times

There is no training as such for walkthrough analysis, and the associated application time is dependent upon the size and complexity of the task or scenario under analysis. The application time for walkthrough analysis is typically very low.

9.6.9 Reliability and Validity

Peute and Jaspers (2007) found that all of the problems that the walkthrough method identified were also present in the think-aloud procedure, providing initial levels of validation support for the method.

9.6.10 Flowchart

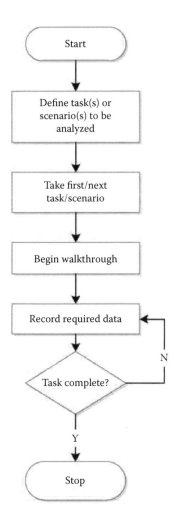

9.6.11 *Tools Needed*

A walkthrough analysis can be conducted using pen and paper. Some form of the concept or system under analysis is also required (e.g., pop-up, prototype, installation, mark out). It is also recommended that video- and audio-recording equipment be used to record the walkthrough.

9.7 Usefulness, Satisfaction, and Ease-of-Use Questionnaire

9.7.1 *Background and Applications*

The usefulness, satisfaction, and ease-of-use (USE) questionnaire was developed by Lund (2001). Lund wanted to create a method that would enable a transferable and comparable *usability score* to be derived for all interfaces. The method aims to incorporate user satisfaction into usability evaluation to utilize users' ability to subjectively evaluate what is usable. Lund argues that current usability methods neglect the user's subjective opinions, whereas USE tries to incorporate these judgments into the development of transferable usability scores. The method involves asking users to rate their agreement, on a seven-point Likert scale, with a set of 30 statements. The 30-statements map onto four high-level categories: usefulness, ease of use, ease of learning, and satisfaction.

9.7.2 *Domain of Application*

The method was originally developed for the evaluation of software, hardware, and services with a core emphasis on its transferability across domains (Lund, 2001).

9.7.3 *Application in Land Use Planning and Urban Design*

Such an approach is central to the LUP & UD necessity for capturing subjective experiences of users in a range of urban systems and settings. Having the ability to empirically explore the insights of a range of users of different urban designs and scenarios is invaluable for both research and practice.

9.7.4 Procedure and Advice

Step 1: Define the task under analysis

The initial stage of any analysis involves defining the task or setting under analysis. This definition guides the analysis and ensures that relevant information is captured. It is important that the analyst presents a clear definition of the goals of the analysis before the next step is conducted.

Step 2: Identify participants

The next stage of this interface evaluation is to define the user population. Once it is clear who the end users are, a representative sample should be recruited to take part in the USE analysis.

Step 3: Design task scenario

At this stage, the analyst should decide and formulate participant instructions regarding which aspects of the environment should be explored and how this exploration will proceed. It is important that participants experience the urban system in the same manner in order for their results to be comparable. The set of tasks should be representative of the system as a whole and must be large enough to represent all aspects of the system, but not so large as to be overwhelming for the participant. It is recommended that an HTA for task under analysis should be used to develop the task list.

Step 4: Briefing

Before the analysis is undertaken, the analyst should ensure that all participants understand what is required of them. It may be useful for the analyst to provide an example of how to complete the questionnaire. At this point, the participants should be encouraged to question anything they are confused or unsure about to ensure that they fully understand the task, the system, and the questionnaire.

Step 5: Task performance

At this stage in the analysis, the participants are asked to undertake the task list in the order that it is listed. They should not be able to help one another or receive assistance from the analyst during this stage. This stage will continue until all participants have completed all tasks on the task list.

Step 6: Complete USE questionnaire

The next stage of the analysis involves the participants completing the USE questionnaire. Once the task list is completed, participants should be given the questionnaire and told to complete all items based on the task list experience with the system under analysis. Again, participants should not be conferring with one another in this stage, although the analyst can aid them in completing the questionnaire.

Step 7: Analyze Data

On completion of the questionnaires, the analyst can begin to score the responses and analyze the data. Calculating average scores may be used for summary analysis.

9.7.5 *Advantages*

- The USE questionnaire provides metrics to reliably measure user reactions to a product.
- It is a very quick method to apply.
- It enables the definition of an acceptable level of user satisfaction (Lund, 1998).
- It provides a baseline evaluation that can measure progress through design iterations (Lund, 2001).
- It allows the identification of priority areas for usability improvement (Lund, 1998).
- It allows insight into design aspects that can increase usability (Lund, 1998).

9.7.6 *Disadvantages*

- Every question in the questionnaire is positively phrased, which means that the results it gives are biased toward positive responding.

9.7.7 *Approximate Training and Application Times*

Both training and application times for the method are low.

9.7.8 *Validity and Reliability*

There seems to be a degree of validity to the method (Lund, 2001), although no data are currently available.

9.7.9 Related Methods

There are many similar usability questionnaires, such as system usability scale (SUS).

9.7.10 Tools Needed

The method can be conducted using pen and paper.

9.7.11 Flowchart

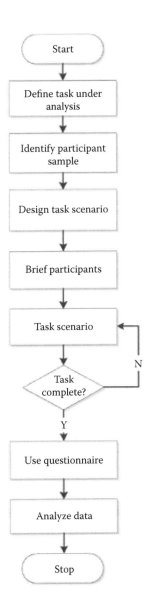

9.7.12 Example

The 30 statements are given in the following (Lund, 1998); each of these is to be rated against a seven-point Likert scale.

Usefulness

1. It helps me be more effective.
2. It helps me be more productive.
3. It is useful.
4. It gives me more control over the activities in my life.
5. It makes the things I want to accomplish easier to get done.
6. It saves me time when I use it.
7. It meets my needs.
8. It does everything I would expect it to do.

Ease of Use

9. It is easy to use.
10. It is simple to use.
11. It is user friendly.
12. It requires the fewest steps possible to accomplish what I want to do with it.
13. It is flexible.
14. Using it is effortless.
15. I can use it without written instructions.
16. I do not notice any inconsistencies as I use it.
17. Both occasional and regular users would like it.
18. I can recover from mistakes quickly and easily.
19. I can use it successfully every time.

Ease of Learning

20. I learned to use it quickly.
21. I easily remember how to use it.
22. It is easy to learn to use it.
23. I quickly became skillful with it.

Satisfaction

24. I am satisfied with it.
25. I would recommend it to a friend.
26. It is fun to use.
27. It works the way I want it to work.
28. It is wonderful.

29. I feel I need to have it.
30. It is pleasant to use.

9.7.13 Recommended Text(s)

Lund, A. M. (2001). Measuring usability with the USE questionnaire12. *Usability Interface, 8*(2), 3–6.

9.8 The System Usability Scale

9.8.1 Background and Applications

SUS offers a very quick and simple to use questionnaire designed to assess the usability of a particular product or urban setting. It consists of 10 usability statements that are rated on a Likert scale of 1 (strongly agree with statement) to 5 (strongly disagree with statement). Answers are coded and a total usability score is derived for the product or device under analysis.

9.8.2 Domain of Application

Generic.

9.8.3 Application in Land Use and Urban Design

Here also SUS provides LUP & UD practitioners and researchers with an empirical way to better understand the usability of artifacts within the urban realm (e.g., wayfinding devices, interpretation displays, seating) or particular urban environments. Further, the generic nature of the questionnaire statements would permit its use to explore the usability of community or stakeholder consultation devices or websites, for example. It also allows for the quick comparison of different designs or indeed the same design from different user perspectives.

9.8.4 Procedure and Advice

Step 1: Create exhaustive task list for the device under analysis
Initially, the analyst should develop an exhaustive task list for the artifact or environment under analysis. This should

include every possible action associated with the artifact or environment. If this is not possible due to time constraints, then the task list should be as representative of the full functionality of the device as possible. An HTA is normally used for this purpose.

Step 2: User trial

Next, the participant should complete a thorough user trial for the object, system, or environment under analysis. They should be instructed to perform every task on the task list given to them.

Step 3: Complete SUS questionnaire

Once the participant has completed the appropriate task list, he or she should be given the SUS questionnaire and instructed to complete it, based on his or her opinions of the device under analysis.

Step 4: Calculate SUS score for the device under analysis

Once completed, the SUS questionnaire score is calculated to derive a usability score for the device under analysis. Scoring an SUS questionnaire is a very simple process. Each item in the SUS scale is given a score between 0 and 4. The items are scored as follows (Stanton and Young, 1999a):

The score for odd numbered items is the scale position, for example, 1, 2, 3, 4, or 5 minus 1.

The score for even-numbered items is 5 minus the associated scale position.

The sum of the scores is then multiplied by 2.5.

The final figure derived represents a usability score for the device under analysis and should range between 0 and 100. The higher the score, the better the usability.

9.8.5 Advantages

- SUS is very easy to use, requiring only minimal training.
- It offers an immediately useful output in the form of a usability *rating*.

- It is very useful for canvassing user opinions of artifacts and environments.
- The scale is generic and so can be applied in any domain.
- It is very useful when comparing two or more devices in terms of usability.
- Its simplicity and speed of use mean that it is a very useful technique to use in conjunction with other usability assessment techniques.
- It is very quick in its application.
- It can be adapted to make it more suitable for other domains.
- It is freely available.

9.8.6 Disadvantages

- The output of the SUS is very limited.
- It requires an operational version of the device or system under analysis.
- It is unsophisticated.
- Non-English speakers have experienced problems understanding the terminology (Finstad, 2010).
- Finstad (2010) questions the adequacy of using a five-point Likert scale, suggesting that a seven-point scale may be more appropriate.
- According to Finstad (2010), the method does not adequately align with the ISO usability standards.

9.8.7 Example

The questionnaire statements for use with the five-point Likert scale for the SUS are given in the following:

1. I think that I would like to use this system frequently.
2. I found the system unnecessarily complex.
3. I thought the system was easy to use.
4. I think that I would need the support of a technical person to be able to use this system.

5. I found the various functions in this system were well integrated.
6. I thought there was too much inconsistency in this system.
7. I would imagine that most people would learn to use this system very quickly.
8. I found the system very cumbersome to use.
9. I felt very confident using the system.
10. I needed to learn a lot of things before I could get going with this system.

10

Systems Analysis and Design Methods

10.1 Introduction

In the current chapter we describe a series of methods, which relate to the so-called systems thinking approach to analysis and design. This is a philosophy currently prevalent within the discipline of human factors, which aims to understand and improve performance and safety in complex sociotechnical systems (STS). It is most prominent in the area of accident analysis and prevention whereby, after first emerging in the early twentieth century (Heinrich, 1931), it is now characterized by a series of accident causation models and analysis methods (Leveson, 2004; Rasmussen, 1997). Contemporary models are underpinned by the notion that safety and accidents are emergent properties arising from nonlinear interactions between multiple components across complex STS (Leveson, 2004). A key focus of systems analysis and design methods is the need to take the overall system as the unit of analysis rather than individual components within the system.

As well as accidents, human factors issues are increasingly being examined through the systems thinking lens (Karsh et al., 2014; Salmon et al., 2017b; Walker et al., 2017). In line with this, since the turn of the century, a range of human factors methods have either been developed or have experienced a resurgence in popularity. These include systems analysis frameworks, such as cognitive work analysis (CWA; Vicente, 1999) and the event analysis of systemic teamwork (EAST; Stanton et al., 2013a); accident analysis methods, such as Accimap (Svedung and Rasmussen, 2002), the systems theoretic accident model and processes (Leveson, 2004), and the functional resonance analysis method (Hollnagel, 2012); and systems design methods, such as the MacroErgonomic Analysis and Design method

(Kleiner, 2006) and the cognitive work analysis (CWA) Design Toolkit (Read et al., 2016).

The aim of the current chapter is to demonstrate how two of these methods can be used to provide in-depth analyses of performance in complex STS, such as those found in the land use planning and urban design (LUP & UD) discipline. The systems thinking approach involves taking the overall system as the unit of analysis, looking beyond individuals, and considering the interactions between humans and between humans and artifacts within a system. This view also encompasses factors within the broader organizational, social, or political system in which behavior takes place. Taking this perspective, behaviors emerge not from the decisions or actions of individuals but from interactions between humans and artifacts across the wider system. At the most basic level when examining LUP & UD within complex STS, the descriptive constructs of interest can be distilled down to simply

- *Why* (the goals of the system, subsystem[s], and actor[s]).
- *Who* (the actors performing the activity are, including humans and technologies).
- *When* (activities take place and which actors are associated with them).
- *Where* (activities and actors are physically located).
- *What* (activities are undertaken, what knowledge/decisions/processes/devices are used and what levels of workload are imposed).
- *How* (activities are performed and how actors communicate and collaborate to achieve goals).

To assist researchers and practitioners in exploring these constructs, two systems analysis methods will be detailed: CWA (Vicente, 1999) and EAST (Stanton et al., 2013a). CWA offers a comprehensive framework for the design, evaluation, and analysis of complex STS. Rather than offering a description of the activity performed within a particular system, the CWA framework provides methods that can be used to develop an in-depth analysis of the constraints that shape agent activity within the system. STS scenarios are often so complex and multifaceted, and analysis requirements so diverse, that various methods need to be applied as one method in isolation cannot cater for the scenario and analysis requirements. Building on a long history and tradition of methods integration in human factors research and practice (Kirwan, 1992; Stanton et al., 2005, 2013a), EAST (Stanton et al., 2008, 2013a) provides an integrated suite of methods for analyzing the performance of complex STS. The framework supports this by providing methods to describe, analyze, and integrate three network-based representations of activity: task, social, and information networks.

10.2 Cognitive Work Analysis

10.2.1 Background and Applications

CWA (Jenkins et al., 2009; Vicente, 1999) is a framework that was developed to support the analysis and design of complex sociotechnical work systems. The framework provides a suite of methods that are used to model different types of constraints, building a model of how work could proceed within a given work system. The focus on constraints separates the technique from other approaches to analysis that aim to describe how work is actually conducted, or prescribe how it should be conducted.

CWA was originally developed at the Risø National Laboratory in Denmark (Rasmussen et al., 1994). In the years that have followed, attempts have been made to add additional detail and clarification to the framework proposed by Rasmussen et al.; however, the underlying framework remains largely unchanged.

The CWA approach can be used to describe the constraints imposed by purpose of a system, its functional properties, the nature of the activities that are conducted, the roles of the different actors, and their cognitive skills and strategies. Rather than offering a prescribed methodology, the CWA framework provides a set of tools that can be used either individually or in combination with one another, depending upon the analysis needs. These tools are divided between phases. The exact names and scopes of these phases differ slightly depending on the scope of the analysis; however, the overall scope remains largely the same. As defined by Vicente (1999), the CWA framework comprises five different phases: work domain analysis (WDA), control task (or activity) analysis, strategies analysis, social organization and cooperation analysis (SOCA), and worker competencies analysis.

The different tools within the CWA framework have been used for various purposes, including system modeling (Chin et al., 1999; McLean et al., 2017; Salmon et al., 2016b; Stanton et al., 2013a), system design (Bisantz et al., 2003; Rasmussen et al., 1994; Read et al., 2017), accident analysis (Stevens and Salmon, 2016), evaluation of design concepts (Cornelissen et al., 2015; Read et al., 2017), urban design and planning (Stevens and Salmon, 2014; Stevens et al., 2017), analyses of decision-making processes (Mulvihill et al., 2016), training needs analysis (Naikar and Sanderson, 1999), design and information requirements specification (Salmon et al., 2016b; Stoner et al., 2003), tender evaluation (Naikar and Sanderson, 2001), team design (Naikar et al., 2003), and error management training design (Naikar and Saunders, 2003). Despite the CWA's origins within the nuclear power domain, the CWA applications referred to earlier have taken place in a wide range of different domains, including naval, military, aviation, road, rail, urban planning, health care, and sport.

10.2.2 Domain of Application

The CWA framework was originally developed for the nuclear power domain; however, the generic nature of the methods within the framework allows it to be applied in a wide range of domains. The framework is best suited to domains that can be described as complex and sociotechnical in nature (i.e., comprising social and technical elements and exuding some or all of the following qualities: dynamic, uncertain, with interconnected parts).

10.2.3 Application in Land Use Planning and Urban Design

CWA is a useful approach to explore complex urban systems. It allows LUP & UD practitioners and researchers the means to better identify the array of interdependencies between the technical and subjective parameters of our cities. CWA provides a way for LUP & UD to reveal the complexity and therein optimize the design of urban environments as STS. To date, the framework has been used for a variety of urban applications including urban corridors (Stevens et al., 2017), pedestrian prioritization (Stevens and Salmon, 2014) and land use and transport integration (Stevens and Salmon, 2016).

10.2.4 Procedure and Advice

It is especially difficult to prescribe a strict procedure for the CWA framework. In its true form, the framework is used to provide a description of different constraint sets within a particular system. This description can then be used to address specific research and design aims. For example, WDA is commonly used to support interface design and evaluation purposes, but it can also be used to inform training design and evaluation. It would also be beyond the scope of this review to describe the procedure fully. The following procedure is intended to act as a broad set of guidelines for each of the phases defined by the CWA framework. A more complete description can be found in Jenkins et al. (2009a) and Stanton et al. (2017a).

Step 1: Determine the Aims and Objectives of the Analysis
   The first step in applying CWA involves clearly defining the tasks or system under analysis along with any analysis boundaries. In addition, the aim of the analysis should be clearly defined.

Step 2: Select Appropriate CWA Phases and Methods
   Defining an appropriate boundary for the analysis is particularly important as this prevents the analysis from becoming too large, complex, and unwieldy. Once the nature and desired outputs of the analysis are clearly defined, the analysis team carefully selects the most appropriate CWA phases and methods to be employed during the analysis. In general, it is recommended that WDA be applied as a starting point as it provides a holistic view of the system. The remaining phases are normally undertaken on the basis of this description; however, it may be that a WDA will suffice. Recent applications of the method have included multiple phases (Salmon et al., 2016b) or WDA alone (Stevens and Salmon, 2016).

   Conduct Steps 3–8 as appropriate.

Step 3: Data Collection
   Once the aims of the analysis are clearly defined and the appropriate phases are chosen, the next step involves collecting targeted data about the system and its behavior. The specific data collected is dependent on the phases being applied; however, data collection for CWA typically involves observations, concurrent verbal protocols, structured or semistructured

interviews (e.g., the Critical Decision Method), questionnaires and surveys, walkthrough analysis and documentation review (e.g., incident reports, standard operating procedures). For studies of expertise that involve applying the work domain analysis, control task analysis, strategies analysis and worker competencies analysis phases it is recommended that the following data collection approaches are used:

- Documentation review and analysis (WDA, SOCA);
- Observations (WDA, Control task analysis, Strategies analysis, SOCA);
- Critical Decision Method interviews with SMEs (Control task analysis, Strategies analysis, SOCA);
- Task walkthroughs with SMEs providing concurrent verbal protocols (Control task analysis, Strategies analysis, Worker competencies analysis);
- SME workshops (WDA, Control task analysis, Strategies analysis, SOCA, Worker competencies analysis).

Step 4: Work Domain Analysis

The initial phase within the CWA framework, WDA, provides a description of the constraints that govern the purpose and the function of the systems under analysis. The abstraction hierarchy (Rasmussen, 1985; Vicente, 1999; see also Figure 10.1) is used to provide a context-independent description of the domain. The abstraction hierarchy consists of five levels of abstraction, ranging from the most abstract level of purposes to the most concrete level of form (Vicente, 1999). The labels used for each of the levels of the hierarchy tend to differ, depending on the aims of the analysis. It is felt that the use of the word *domain* in the top three levels and the use of the word *physical* in the bottom two levels draw a fitting distinction:

- *Domain purpose*: The domain purpose, displayed at the very top of the diagram, represents the reason why the work system exists. This purpose is independent of any specific situation and is also independent of time—the system purpose exists as long as the system does.
- *Domain values*: The domain values level of the hierarchy is used to capture the key values that can be used to

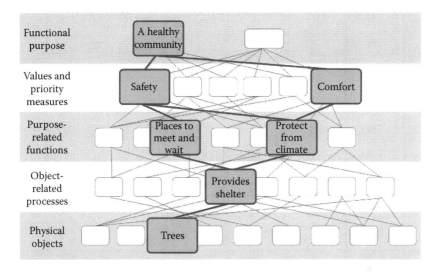

Figure 10.1   The five levels of the abstraction hierarchy. (Adapted from Vicente, 1999, in Stevens, N. J. et al., Work domain analysis applications in urban planning: Active transport infrastructure and urban corridors, in *Cognitive Work Analysis: Applications, Extensions and the Future*, Taylor & Francis Group, Boca Raton, FL, 2017.)

assess how well the work system is performing its domain purpose(s). These values are likely to be conflicting.

- *Domain functions*: The middle layer of hierarchy lists the functions that can be performed by the combined work system. These functions are expressed in terms of the domain in question.
- *Physical functions*: The physical functions that the objects can perform are listed. These are listed generically and are independent of the domain purpose.
- *Physical objects*: The key physical objects within the work system are listed at the base of the hierarchy. These objects represent the sum of the relevant objects from all of the component technologies. This level of the diagram is independent of purpose; however, analyst judgment is required to limit the object list to a manageable size.

The structure of the abstraction hierarchy framework acts as a guide to acquiring the knowledge necessary to understand the domain. The framework helps to direct the search for deep knowledge about the work domain, providing structure to the document analysis process, particularly for the domain novice. Although the output may initially appear overbearing, its value to the analysis cannot be overstated. The abstraction hierarchy defines the systemic constraints at the highest level.

The top three levels of the diagrams consider the overall objectives of the domain, and what it can achieve, whereas the bottom two levels concentrate on the physical components and their affordances. Through a series of *means–ends* links, Figure 10.2, it is possible to model the *what*, *why*, and *how* individual components can have an impact on the overall domain purpose.

The abstraction hierarchy is constructed by considering the work system's objectives (top–down) and the work system's capabilities (bottom–up). The diagram is constructed on the basis of a range of data collection opportunities. The exact data collection procedure is dependent on the domain in question and the availability of data. In most cases, the procedure commences with some form of document analysis. Document analysis allows the analyst to gain a basic domain understanding, forming the basis for semistructured interviews with domain experts. Wherever possible, observation of the urban context is highly recommended. Table 10.1 provides a series of prompts for inclusions at each of the WDA hierarchy levels.

Step 5: Control Task Analysis (Contextual Activity Template)

Up until this point, the analysis has deliberately not considered the constraints that are imposed by specific situations. Control task analysis shifts the focus onto the constraints that are imposed by specific situations. One tool for considering such constraints is the *contextual activity template* (CAT; Naikar et al., 2006). This tool plots the functions, affordances, and objects identified in the abstraction hierarchy against a number of specific *situations*. At this stage, the analysis remains independent of the actor. The first stage of the process is to define the situations. Situations can be characterized by either time or location (or a combination of the two). In many cases, it is appropriate to explore more than one set of situations using multiple CAT representations to meet a range of analytic goals.

Next, depending on the focus of the analysis, nodes from the relevant level of the WDA are added to the CAT. For

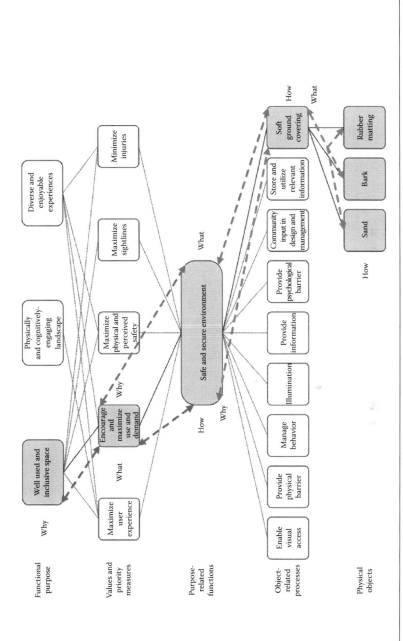

Figure 10.2 Example of means–ends relationships within a playground design. (From Missen, L. et al., A sociotechnical systems analysis approach to playground design, in *Contemporary Ergonomics 2017, Proceedings of the Chartered Institute of Ergonomics and Human Factors Annual Conference*, April, Daventry, UK, 2017.)

Table 10.1 Example WDA Prompts

	PROMPTS	KEYWORDS
Functional purposes	Purposes • For what reasons does the work system exist? • What are the highest level objectives or ultimate purposes of the work system? • What role does the work system play in the environment? • What has the work system been designed to achieve? External constraints • What kinds of constraints does the environment impose on the work system? • What values does the environment impose on the work system? • What laws and regulations does the environment impose on the work system?	Purposes Reasons, goals, objectives, aims, intentions, mission, ambitions, plans, services, products, roles, targets, aspirations, desires, motives, values, beliefs, views, rationale, philosophy, policies, norms, conventions, attitudes, customs, ethics, morals, principles. External constraints Laws, regulations, guidance, standards, directives, requirements, rules, limits, public opinion, policies, values, beliefs, views, rationale, philosophy, norms, conventions, attitudes, customs, ethics, morals, principles.
Values and priority measures	• What criteria can be used to judge whether the work system is achieving its purposes? • What criteria can be used to judge whether the work system is satisfying its external constraints? • What criteria can be used to compare the results or effects of the purpose-related functions on the functional purposes? What are the performance requirements of various functions in the work system? How is the performance of various functions in the work system measured or evaluated and compared?	Criteria, measures, benchmarks, tests, assessments, appraisals, calculations, evaluations, estimations, judgments, scales, yardsticks, budgets, schedules, outcomes, results, targets, figures, limits. Measures of effectiveness, efficiency, reliability, risk, resources, time, quality, quantity, probability, economy, consistency, frequency, success. Values—laws, regulations, guidance, standards, directives, requirements, rules, limits, public opinion, policies, values, beliefs, views, rationale, philosophy, norms, conventions, attitudes, customs, ethics, morals, principles.

(Continued)

Table 10.1 (*Continued*)   Example WDA Prompts

	PROMPTS	KEYWORDS
Purpose-related functions	• What functions are required to achieve the purposes of the work system? • What functions are required to satisfy the external constraints on the work system? • What functions are performed in the work system? • What are the functions of individuals, teams, and departments in the work system?	Functions, roles, responsibilities, purposes, tasks, jobs, duties, occupations, positions, activities, operations.
Object-related processes	• What can the physical objects in the work system do or afford? • What processes are the physical objects in the work system used for? • What are the functional capabilities and limitations of physical objects in the work system?	Processes, functions, purposes, utility, role, uses, applications, functionality, characteristics, capabilities, limitations, capacity, physical processes, mechanical processes, electrical processes, chemical processes.
Physical objects	• What are the physical objects or physical resources in the work system—both man-made and natural? • What physical objects or physical resources are necessary to enable the processes and functions of the work system?	Human-made and natural objects—tools, equipment, devices, apparatus, machinery, items, instruments, accessories, appliances, implements, technology, supplies, kit, gear, buildings, facilities, premises, infrastructure, fixtures, fittings, assets, resources, staff, people, personnel, terrain, land, meteorological features.

Source: Naikar, N. et al., *Work Domain Analysis: Theoretical Concepts and Methodology*. Defence Science & Technology Organisation Report, DSTO-TR-1665, Melbourne, Australia, 2005.

example, if the analysis is focused on object-related constraints, the physical objects level of the WDA should be added to the CAT. On the other hand, if the analysis is focused on functions, then the generalized processes level of the WDA should be added.

Once the CAT matrix is developed, analysts should work through the matrix and specify, for each object/affordance/function, where they currently reside or are undertaken, and where they could reside or be undertaken given modifications of constraints (i.e., through system redesign).

Thus, the products provide a more context-specific description of the domain. The CATs can be used to inform information-exchange requirements by indicating situations in which information may be required. Likewise, by adding in additional constraints, the list of possible information requirements is reduced in certain situations.

Step 6: Control Task Analysis (Decision Ladders)

Continuing with the theme of describing additional constraint, key function-situation cells within the CAT can be explored in terms of decision-making. The decision ladder (Rasmussen, 1974; Figure 10.2) is the tool most commonly used within CWA to describe decision-making activity. Its focus is on the entire decision-making activity rather than the moment of selection between options. It is not specific to any single actor; instead, it represents the decision-making process of the combined urban system. In many cases, the decision-making process may be collaborative, distributed between a range of human and technical decision-makers.

The ladder contains two different types of node: the rectangular boxes represent data processing activities and the circles represent resultant states of knowledge. Novice users (to the situation) are expected to follow the decision ladder in a linear fashion, whereas expert users are expected to link the two halves by shortcuts known as leaps and shunts. Leaps connect two states of knowledge (circle to circle on the decision ladder) and shunts connect an information processing activity to a state of knowledge (box to circle on the decision ladder). The left side of the decision ladder represents the observation of the current system state, whereas the right side represents the planning and execution of tasks and procedures to achieve a target system state. Decision ladders can be populated on the basis of semistructured interviews with subject matter experts (SMEs) (Jenkins et al., 2008). This involves coding the interview data to each part of the decision ladder (e.g., alert, information, system state, goals, and options) and also using the interview data to identify instances of leaps and shunts. Ostensibly, in its raw form, the decision-ladder models provide an overview of the decision-making process under analysis along with a list of the information requirements for making a decision triggered by a number of presupposed events. At this stage of the analysis, the relative importance of these information elements is not considered (Figure 10.3).

Step 7: Conduct Strategies Analysis

The aim of strategies analysis phase is to describe the constraints that dictate how a system can be moved from one state to another. The strategies analysis phase therefore involves identifying the different strategies that can potentially be employed to achieve control tasks; as Jenkins (2009) pointed out, the strategy adopted in a given situation may vary significantly depending upon the constraints imposed. Naikar et al. (2006) suggested that strategies analysis is concerned with identifying general categories of cognitive procedures. The strategy adopted is dependent upon various factors, including workload and task demands, training, time pressure, experience, and familiarity with the current situation. Ahlstrom (2005) proposed a modified form of Vicente's (1999) information flow map for the strategies analysis phase; here the situation is broken down into a *start state* and a desired *end state*. The strategies available for achieving the end state connect the two states. An example of a simplified flow map is presented in Figure 10.4. The strategies available are typically identified via interviews with appropriate SMEs.

Although information flow maps are typically used for the strategies analysis component of CWA, other tools,

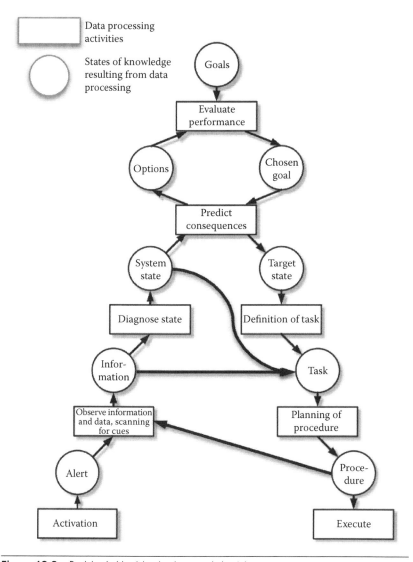

Figure 10.3   Decision ladder (showing leaps and shunts).

such as the strategies analysis diagram (Cornelissen et al., 2013), can also be applied. The strategies analysis diagram builds on the abstraction hierarchy developed in the WDA phase and involves the addition of two levels to the diagram: verbs and criteria. The verbs are used to specify how the physical objects can be used. The criteria are then used to specify the circumstances under which different strategies might be chosen.

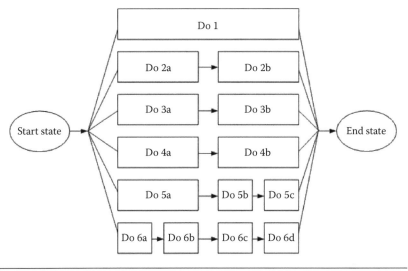

Figure 10.4 Strategies analysis simplified flow map.

Step 8: Conduct Social Organization and Cooperation Analysis

The SOCA phase involves identifying how activities and strategies are and can be distributed between agents (human and nonhuman) within the system. The objective is to determine how social and technical factors can work together in a way that maximizes system performance. SOCA is undertaken on outputs from the other CWA phases, such as the abstraction hierarchy, CAT, decision ladders, and information flow maps. This involves shading nodes to show who or what can undertake functions, provide affordances, engage in parts of the decision-making process, and complete different components of strategies. This provides a systemic description that can be used to explore optimal allocation of functions across the system.

As an example, when applying SOCA to the abstraction hierarchy, the first step involves identifying all the relevant actors within the system and allocating a color to each. This includes human (e.g., driver, pedestrian, and cyclist) and nonhuman actors (e.g., legislation, plans, and policy). Next the analyst works through each of the nodes in the abstraction hierarchy and determines which actors currently perform

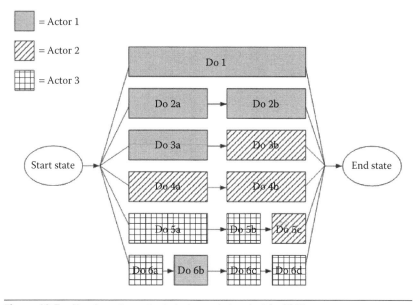

Figure 10.5   Mapping actors onto strategies analysis output for SOCA.

or provide each node, followed by identifying which actors could perform or provide each given design modifications.

An example of SOCA, based on the strategies analysis flow map presented earlier, is depicted in Figure 10.5. SOCA is best informed through SME interviews, observation, and review of relevant documentation such as training manuals and standard operating procedures.

Step 9: Conduct Worker Competencies Analysis

The final phase, WCA, involves identifying the competencies required for undertaking activity within the system in question. This phase is concerned with identifying the psychological constraints that are applicable to system design (Kilgore and St-Cyr, 2006). Vicente (1999) recommends that the skill, rule, and knowledge (SRK) framework (Rasmussen, 1983; cited in Vicente, 1999) be used for this phase. The SRK framework describes three hierarchical levels of human behavior: skill, rule, and knowledge-based behavior. Each of the levels within the SRK framework defines a different level of cognitive control or human action (Vicente, 1999).

Skill-based behavior occurs in routine situations that require highly practiced and automatic behavior and in which there is only small conscious control on behalf of the operator. According to Vicente (1999), skill-based behavior consists of smooth, automated, and highly integrated patterns of action that are performed without conscious attention. The second level of behavior, the rule-based level, occurs when the situation deviates from the normal but can be dealt with by the operator applying rules that are either stored in memory or are readily available; for example, emergency procedures. According to Vicente (1999), rule-based behavior consists of stored rules derived from procedures, experience, instruction, or previous problem-solving activities. The third and highest level of behavior is knowledge-based behavior, which typically occurs in nonroutine situations (i.e., emergency scenarios) in which the operator has no known rules to apply and has to use problem solving skills and knowledge of the system characteristics and mechanics to achieve task performance. According to Vicente (1999), knowledge-based behavior consists of deliberate, serial, search based on an explicit representation of the goal and a mental model of the functional properties of the environment. Further, knowledge-based behavior is slow, serial, and effortful, as it requires conscious, focused attention (Vicente, 1999). The SRK framework is presented in Figure 10.6.

Step 10: SME Review

Once the initial draft analyses for each phase are complete it is useful to have various SMEs review it. The analyses should then be refined based on SME feedback. be It is normal practice for the CWA outputs to go through many iterations before they are finalized.

10.2.5 Example

Stevens and Salmon (2014) used CWA to describe the engineering and urban design relationships of footpaths. The aim of this study was to use a systems analysis and design framework to develop a design

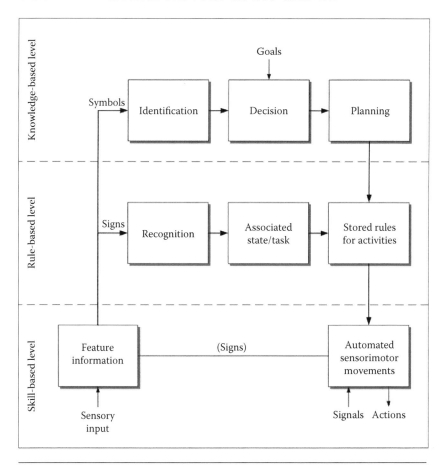

Figure 10.6   SRK behavioral classification scheme.

template for an *ideal* footpath system that embodies both safety and sense of place. This was achieved through using the first phase of the CWA framework, WDA shown in full in Figure 10.7 (summarized in Figure 10.8), to specify a model of footpaths as safe places for pedestrians.

10.2.6 Advantages

- CWA offers a comprehensive framework for the design and analysis of complex systems.
- The CWA framework is based on sound underpinning theory, is extremely flexible, and can be applied for a number of different purposes.

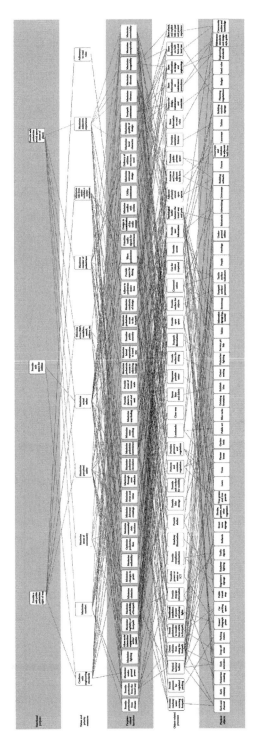

Figure 10.7   WDA of footpath system. (From Stevens, N. and Salmon, P., *Accid. Anal. Prev.*, 72, 257–266, 2014.)

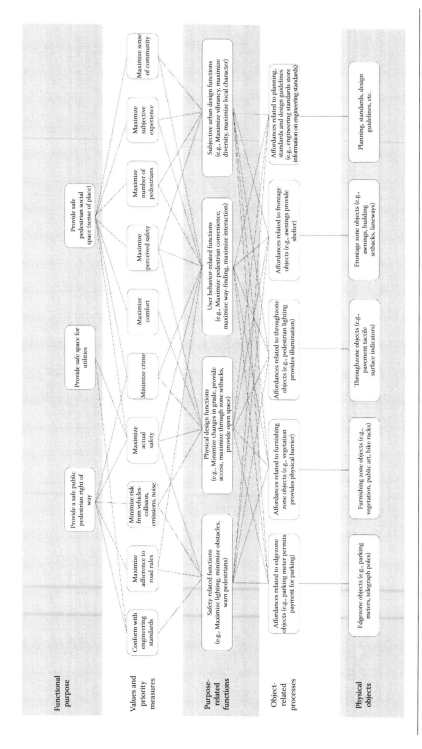

Figure 10.8   Summarized WDA of footpath system. (From Stevens, N. and Salmon, P., *Accid. Anal. Prev., 72*, 257–266, 2014.)

- The diversity of the different methods within the framework enables a comprehensive analysis.
- The methods within the framework are extremely useful. The abstraction–decomposition space in particular can be used for a wide range of purposes.
- It can be applied in a number of different domains.
- CWA supports both analysis and design. Indeed its formative nature has seen it employed for system design purposes across many domains (Read et al., 2017).
- CWA now has an associated system design tool, the CWA-Design Toolkit (Read et al., 2016, 2017).

10.2.7 Disadvantages

- The methods within the framework are complex and practitioners may require considerable training in their application.
- A full five phase CWA analysis requires significant data collection activities which can be time-consuming, costly, and difficult to arrange.
- The CWA methods are initially time-consuming to apply (although the products can be reused).
- Some of the methods within the framework are still in their infancy and there is limited prescriptive guidance available on their usage.
- The reliability of the methods may be questionable.
- CWA outputs can be large, unwieldy, and difficult to communicate to practitioners.

10.2.8 Related Methods

The CWA approach does not explicitly define the methods for each of the different CWA phases. Vicente (1999) described the following approaches for the CWA framework: the abstraction–decomposition space (WDA), decision ladders (control task analysis), information flow maps (strategies analysis), and the SRK framework (worker competencies analysis). More recently, Cornelissen et al. (2013)

developed the strategies analysis diagram. For system design purposes, Read et al. (2016) recently developed the CWA-Design
Toolkit.

Various data collection methods have also been used to support
CWAs. These include semistructured interviews, concurrent verbal
protocols, observation, surveys and questionnaires, and documentation review.

10.2.9 Approximate Training and Application Times

The methods used within the CWA framework are complex and
there is also limited practical guidance available on their application.
The training time associated with the CWA framework is therefore
high, particularly if all phases of the framework are to be undertaken.
Due to the exhaustive nature of the CWA framework and the methods used, the application time is also considerable. The time taken
to complete an analysis will be dependent on the size of the system
and the analysis objectives. Naikar and Sanderson (2001) reported
that a WDA of the airborne early warning and control system took
six months to complete. Conversely, smaller-scale studies can be conducted in a few weeks (Stevens and Salmon, 2014).

10.2.10 Reliability and Validity

The reliability and validity of the CWA framework are difficult to
assess. The flexibility and diversity of the methods used ensure that
reliability is impossible to address, although it is apparent that the
reliability of the approaches used may be questionable.

10.2.11 Tools Needed

Most methods within the CWA framework can be applied using
pen and paper only. However, typically interviews and observations
are required, and so audio- and video-recording equipment may be
required. CWA outputs are also typically large and require software
support in their construction. For example, Microsoft Visio is particularly useful in the construction of abstraction hierarchies, decision
ladders, strategies analysis diagrams, and so on.

10.2.12 Flowchart

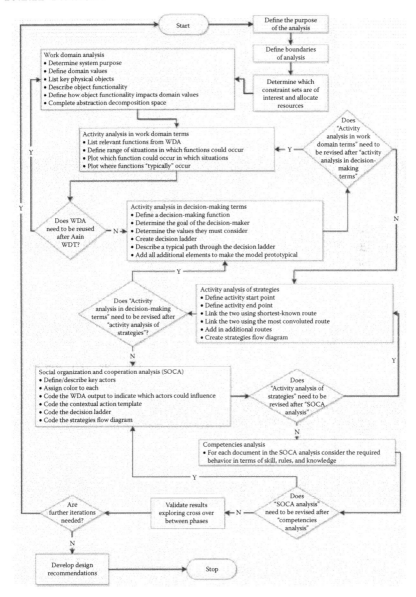

10.3 Event Analysis of the Systemic Teamwork

10.3.1 Background and Applications

The EAST (Stanton et al., 2008) provides a framework of methods that allows system performance to be comprehensively described and evaluated. Since its conception, the framework has been applied in many domains, including land and naval warfare (Stanton et al., 2006a; Stanton et al., 2014a), aviation (Stewart et al., 2008), air traffic control (Walker et al., 2010), road transport (Salmon et al., 2014a) the emergency services (Houghton et al., 2008), and elite cycling (Salmon et al., 2017a).

Underpinning the approach is the notions that distributed team-work can be meaningfully described via a *network of networks* approach in Figure 10.9. Specifically, three networks are considered: task, social, and information networks. Task networks describe the goals and subsequent tasks being performed within the system. Social networks analyze the organization of the system (i.e., communications structure) and the communications taking place between the actors working in the team. Finally, information networks describe the information and knowledge (situation awareness) that the different actors use and share during task performance.

10.3.2 Domain of Application

EAST is a generic approach that was developed specifically for the analysis of teamwork in STS. As such it can be used in any domain in which social and technical elements are working together in pursuit of a common goal.

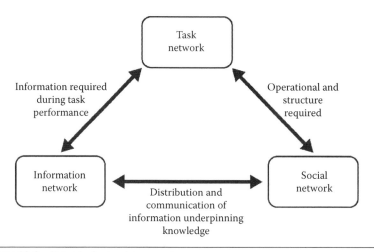

Figure 10.9   Network of networks approach.

10.3.3 Application in Land Use Planning and Urban Design

LUP & UD is most often a product of multidisciplinary approaches to complex environments and STS. Our projects and processes require different resources from a variety of participants over a range of time frames. Although we may not always define these approaches as teamwork, the EAST approach has the potential to offer critical insights into more effective and efficient cooperative processes and project performance.

10.3.4 Procedure and Advice

Step 1: Define Analysis Aims

First, the aims of the analysis should be clearly defined so that appropriate scenarios are used and relevant data are collected. In addition, not all components of the EAST framework may be required, so it is important to clearly define the aims at this point to ensure that the appropriate EAST methods are applied.

Step 2: Define the Task/System Under Analysis

Next, the task (or tasks) or scenario (or scenarios) under analysis should be clearly defined. This is dependent upon the aims of the analysis and may include a range of tasks or one task in particular. It is normally standard practice to develop a hierarchical task analysis (HTA) for the task under analysis if sufficient data and SME access are available. This is useful later on in the analysis and is also enlightening, allowing the analyst to gain an understanding of the task before the observation and analysis begins.

Step 3: Data Collection

Once the aims of the analysis are clearly defined, the next step involves collecting targeted data about the system and its behavior. The specific data collected are dependent on the analysis aims and the resources available; however, data collection for EAST typically involves observations, concurrent verbal protocols, structured or semistructured interviews (e.g., the critical decision method [CDM]), walkthrough analysis, and documentation review (e.g., incident reports, standard operating procedures).

The observation step is often the most important part of the EAST procedure. Typically, a number of analysts are used to observe the system or scenario under analysis. All activities involved in the scenario under analysis should be recorded along an incident timeline, including a description of the activity undertaken, the agents involved, any communications made between agents, and the technology involved. Additional notes should be made where required, including the purpose of the activity observed, any tools, documents, or instructions used to support activity, the outcomes of activities, any errors made, and also any information that the agent involved feels is relevant. In addition, it is useful to video record the task and record verbal transcripts of all communications if possible.

Once the task under analysis is complete, each *key* agent (e.g., scenario commander, agents performing critical tasks) involved should be subjected to a CDM interview. This involves dividing the scenario into key incident phases and then interviewing the actor involved in each phase using a set of predefined CDM probes (O'Hare et al., 2000; see also Chapter 4 for more information on the CDM).

Step 4: Transcribe Data

Once all of the data are collected, it should be transcribed to make it compatible with the EAST analysis phase. An event transcript should then be constructed. This should describe the scenario over a timeline, including descriptions of activity, the actors involved, any communications made, and the technology used. To ensure the validity of the data, the scenario transcript should be reviewed by one of the SMEs involved.

Step 5: Construct Task Network

The first analysis step involves constructing a task network. Prior to this, the initial HTA should be reviewed and refined on the basis of the data collected during Step 3. The data-transcription process allows the analyst to gain a deeper and more accurate understanding of the scenario under investigation. It also allows any discrepancies between the initial HTA scenario description and the actual activity observed to be resolved. Typically, activities in complex sociotechnical systems do not run entirely according to protocol, and certain tasks may have been performed during the scenario that were not described in the initial HTA description. The analyst should compare the scenario transcript with the initial HTA and add any changes as required.

Constructing the task network involves identifying high-level tasks and the relationships between them and creating a network to represent this. Some general rules around the construction of EAST networks are presented in Table 10.2.

Step 6: Conduct Social Network Analysis

A social network analysis (SNA; Driskell and Mullen, 2004) is used to analyze the relationships (e.g., communications, transactions) between the agents involved in the scenario under analysis. This involves first creating a social network matrix should the relationships between agents following by a social network diagram that provides a visual representation of the social network. Typically, the direction (i.e., from actor A to actor B) frequency, type, and content of associations are recorded. It is normally useful to conduct a series of SNAs representing different phases of the task under analysis (using the task phases defined during the CDM part of the analysis).

Step 7: Construct Information Networks

The final step of the EAST analysis involves constructing information networks (see Chapter 7 for a full description) for each scenario phase identified during the CDM interviews. Following construction, information usage should be defined for each actor involved via shading of the information elements within the propositional networks.

Step 8: Construct Composite Networks

Composite networks are used to explore the relationships between tasks, agents, and information (Stanton, 2014). As such, composite networks are constructed by combining the different networks. For example, a task by agents network can be constructed by combining the task and social network to show which tasks are undertaken by which agents. This involves assigning a color to the different agents within the social network and shading each node within the task

Table 10.2 Analysis Rules Regarding the Relationships between Nodes within EAST Networks

NETWORK	NODES	RELATIONSHIPS	EXAMPLES
Task network	Represent high-level tasks that are required during the scenario under analysis. High level tasks are typically extracted from the subordinate goals level of the HTA	Represent instances where the conduct of one high-level grouping of tasks (i.e., task network node) influences, is undertaken in combination with, or is dependent on another group of tasks	The nodes *Identify legal constraints* and *Identify site and zoning* are linked because the zoning cannot be established until the site has been legally identified.
Social network	Represent human, technological, or organizational agents who undertake one or more of the tasks involved in the scenario under analysis (as identified in the HTA and task network)	Represent instances where agents within the social network interact with one another during the scenario under analysis	The nodes *Urban planner* and *community* are linked as the planner needs to communicate with and understand the local community if an informed analysis of the site is to be established.
Information network	Represent grouped categories of information that are required by agents when undertaking scenario under analysis (as identified in the task and social network)	Represent instances where information influences other information or is used in combination with other information in the network during the scenario under analysis	The nodes *views* and *topography* are linked as the establishment of views requires appropriate topography.

network to show which agent performs that particular task. Useful composite networks to construct include

- Task by agents network (combined task and social network)
- Information by agents network (combined information and social network)
- Task and associated information network (combined task and information network)
- Information by agents and tasks network (combined task, social, and information network)

Once the EAST networks are complete, it is pertinent to validate the outputs using appropriate SMEs and recordings of the scenario under analysis. Any problems identified should be corrected at this point.

Step 9: Analyze Networks

An important component of EAST analyses involved using network metrics to analyze the task, social, and information networks. This enables analysis of the structure of the networks and identification of key nodes (e.g., tasks, agents, information) within the networks. Three popular network analysis metrics that have previously been used to interrogate EAST networks include

1. *Network density (overall network)*: Network density represents the level of interconnectivity of the network in terms of relations between nodes. Density is expressed as a value between 0 and 1, with 0 representing a network with no connections between nodes, and 1 representing a network in which every node is connected to every other node. Higher density values are indicative of a well-connected network in which tasks, agents, information, and controls are tightly coupled.

2. *Sociometric status (individual nodes)*: Sociometric status provides a measure of how *busy* a node is relative to the total number of nodes within the network under analysis (Houghton et al., 2006). In the present analysis, nodes with sociometric status values greater than the mean sociometric status value plus one standard deviation are taken to be *key* (i.e., most connected) nodes within each network. These nodes represent either key tasks, agents, pieces of

information, or controls. For example, in the case of the social network, the node with the highest sociometric status is the agent that is the most interrelated with other agents based on communication.

3. *Centrality* (*individual nodes*): Centrality is used to examine the standing of a node within a network based on its geodesic distance from all other nodes in the network (Houghton et al., 2006). Central nodes represent those that are closer to the other nodes in the network as, for example, information passed from one to another node in the network would travel through less nodes. Houghton et al. (2006) point out that well-connected nodes can still achieve low centrality values as they may be on the periphery of the network. For example, in the case of the social network, nodes with higher centrality status values are those that are closest to all other agents in the network as they have direct rather than indirect links with them.

10.3.5 *Advantages*

- The analysis produced is extremely comprehensive, and activities are analyzed from various perspectives.
- The analysis is both qualitative (networks) and quantitative in nature (network analysis metrics).
- Composite networks enable analysts to explore the relationships between tasks, agents, and information.
- The use of network analysis metrics enables analysts to identify key tasks, agents, and information.
- The framework can be used both retrospectively and predictively to forecast system behavior (Stanton and Harvey, 2017).
- The framework approach allows methods to be chosen on the basis of analysis requirements.
- EAST has been applied in a wide range of different domains for various purposes.
- The approach is generic and can be used to evaluate activities in any domain.

- A number of HF concepts are evaluated, including distributed situation awareness, cognition, decision-making, teamwork, and communications.
- It uses structured and valid HF methods and has a sound theoretical underpinning.
- It has great potential to be applied in LUP & UD context to explore multidisciplinary performance.

10.3.6 Disadvantages

- When undertaken in full, the EAST framework is a very time-consuming approach.
- The use of various methods ensures that the framework incurs a high training time.
- To conduct an EAST analysis properly, a high level of access to the domain, task, and SMEs is required.
- Some parts of the analysis can become overly time-consuming and laborious to complete.
- Some of the outputs can be large, unwieldy, and difficult to present in reports, papers, and presentations.
- Reliability and validity have not yet been formally tested.

10.3.7 Related Methods

EAST uses HTA, SNA, and information networks.

10.3.8 Approximate Training and Application Times

Due to the number of different methods involved, the training time associated with the EAST framework is high. Similarly, application time is typically high, although this is dependent upon the task under analysis and the scope of the analysis.

10.3.9 Reliability and Validity

Due the number of different methods involved, the reliability and validity of the EAST methods is difficulty to assess. Indeed, it has not yet been formally tested.

10.3.10 Flowchart

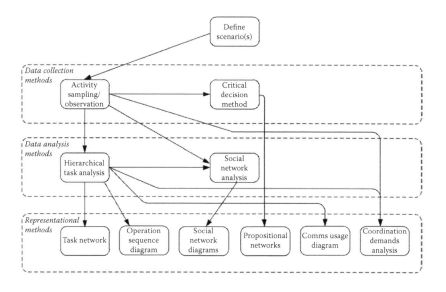

10.3.11 Tools Needed

Normally, video- and audio-recording devices are used to record the activities under analysis. A drawing software package such as Microsoft Visio is also typically used to reproduce the networks. The Agna SNA software tool is typically used to quantitatively analyze the networks.

10.3.12 Example

In the following example, EAST was used to examine a generic site-analysis process. Task, social, and information networks were constructed to describe the key tasks, agents, and information used during site analysis.

Initially, a task network was constructed on the basis of an HTA of a generic site-analysis process (Figure 10.10).

As shown in Figure 10.10, 11 key interrelated tasks were identified. The task network is a dense one with many interdependencies between tasks, suggesting that the tasks required are tightly coupled. In particular, the tasks of analyzing the neighborhood and determining circulation patterns are the most connected within the task network, suggesting that they are central to the overall site-analysis process.

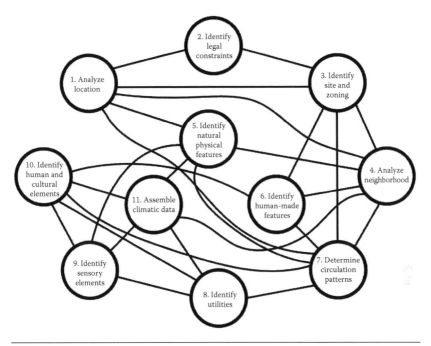

Figure 10.10   Task network of the generic site analysis.

A social network diagram was constructed on the basis of identifying which agents are required to communicate with each other during the site-analysis process (Figure 10.11). The social network demonstrates that there are 19 distinct agents involved in the site-analysis process. This reliance on multiple actors to assemble the required information for the site analyses is central to the work within LUP & UD disciplines. In contrast to the well-connected task network, however, there are few connections between the agents within the social network, suggesting that the network of agents involved in site analysis is loosely coupled.

The social network suggests that, although many actors are necessary for the assembly of the required information, they are most often working independently, whereas a central organizational agent coordinates their responses. This central agent is the urban planner and designer, with the social network diagram revealing that most connected agents within the site-analysis process are urban planners and designers, with connections to all other agents involved in the process. Indeed, urban planners and designers are the only agent in the social network to have connections with more than four other agents. This indicates that urban planners and designers are the key

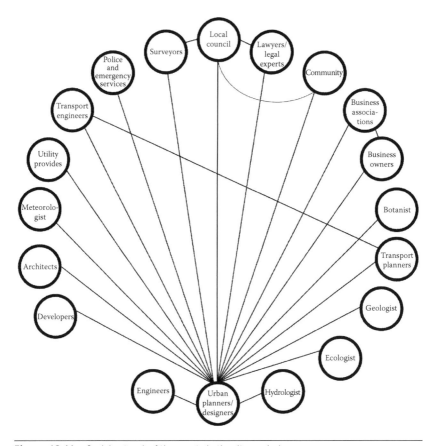

Figure 10.11   Social network of the agents in the site-analysis process.

agents within the site-analysis process, with the local council being the next most connected. The structure of the social network diagram suggests that there may be some simplistic interventions that could improve the site-analysis process. Logical interventions would be to attempt to increase the connectivity of the network through incorporating a requirement for further communication between agents and to reduce the load placed on urban planners and designers.

An information network showing the information required during the site-analysis process was constructed on the basis of the HTA (Figure 10.12). According to the information network, multiple sources of information are required to complete the site-analysis process ranging from information on locations, topography, views, drainage, climate and utilities to traffic, urban form, sensory elements and commercial, retail, residential, and community functions.

Figure 10.12   The information network for the site-analysis task.

Examining the connectedness of different nodes within the network suggests that there are various critical pieces of information required including locations, commercial functions, neighborhood context, road hierarchy, residential functions, retail functions, pathways, and travel times.

To demonstrate the composite network function of EAST, the information and social networks of three tasks were combined. These tasks are 'identify human and cultural elements', 'identify sensory elements', and 'identify natural physical elements'. Figure 10.13 presents this composite network—these representations are most often presented in color, however in this image the actors are numerically identified.

For each task the required information is presented, including the actors responsible for that information and the links between different

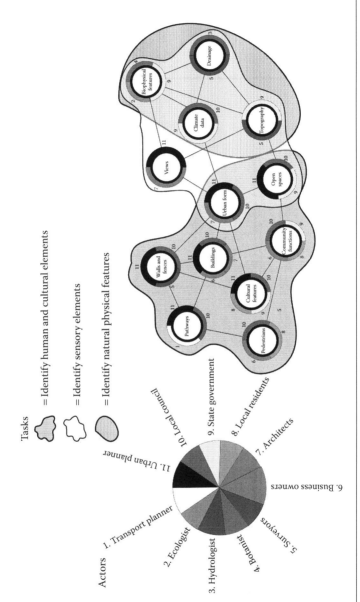

Figure 10.13 Composite network of the information and actors required for three site analysis tasks.

information components. When multiple tasks are considered it is then possible to identify information that is shared between them. Such insights make the interdependencies of the actors, tasks and information required LUP & UD explicit, providing powerful interpretations of urban complexity. This example reveals that all information nodes are 'shared' —there is not one piece of information that only one actor either uses or provides. This composite network also identifies that 'urban form' is the key information node with the most connections. While the importance of elucidating urban form in site analyses is intuitively understood in LUP & UD, EAST permits the empirical identification of the crucial role it plays in a range of tasks undertaken by a variety of actors. The critical role of the urban planner and designer is also apparent with their identification in the majority of information nodes. This highlights a key responsibility in perhaps organizing or facilitating a broader site analysis process for an urban development project.

10.3.13 Recommended Text(s)

Baber, C., Harris, D., & Stanton, N. A. (2012). *Modelling command and control: Event analysis of systemic teamwork.* Farnham: Ashgate Publishing.
Salmon, P. M., Lenne, M. G., Walker, G. H., Stanton, N. A., & Filtness, A. (2014). Using the event analysis of systemic teamwork (EAST) to explore conflicts between different road user groups when making right hand turns at urban intersections. *Ergonomics, 57*(11), 1628–1642.

10.4 ActorMap Analysis

The Accimap technique has been detailed in Chapter 6 (Rasmussen, 1997; Svedung and Rasmussen, 2002) as an approach used to identify and represent the causal flow of events upstream from an accident. It also explores the planning, management, and regulatory bodies that may have contributed to the accident (Svedung and Rasmussen, 2002). As such, a key component of developing the Accimap framework is the ActorMap (Step 2), and in this chapter, we are exploring ActorMaps as a standalone approach for better understanding complex urban systems. ActorMaps also utilize the six levels of (1) government policy and budgeting, (2) regulatory bodies and associations, (3) local area government planning and budgeting, (4) technical and operational management, (5) physical processes and actor activities, and (6) equipment and surroundings.

When considering complex multistakeholder systems, such as those found in the LUP & UD disciplines, this approach allows for clearer interpretations of the following: (1) which actors are part of, or have a stake in, the system (whether they recognize it or not); and (2) the level of interaction between these actors, either by way of formal avenues such as legislative responsibility, or indeed informal connections, perhaps by way of spatial or geographical proximity. Actors are identified at each of the levels and linked between and across the levels based on their relations.

10.4.1 Domain of Application

ActorMap analysis is a generic approach, which has been utilized largely as a component of the Accimap process. However, ActorMaps rely less on retrospective data than Accimaps and can be used to explore current systems of actors.

10.4.2 Application in Land Use Planning and Urban Design

For LUP & UD, it may be used as a valuable means to establish a clear understanding of the range and interactions of all actors within a particular setting or development scenario. Further, it may be used to undertake stakeholder and policy analysis of urban development and planning stakeholders and identify strategic relations.

10.4.3 Procedure and Advice

Step 1: Data Collection

The first step involves clearly identifying the system you are exploring; it therefore involves collecting data regarding the region, policy, or urban-development scenario in question. Data collection for Accimaps can involve a range of activities, including interviews with those involved in the system of interest or SMEs for the domain in question, analyzing reports, legislation, and strategic documents, and observing the scenario or system.

Step 2: Identify Actors across the Levels

Utilizing the six levels as a guide, the analyst should begin to explore and identify a draft list of key actors at those levels involved in the scenario in question. They should also make notes regarding any strategic documents or policies associated with that actor, including links to additional actors.

Step 3: Establish Links between Actors

After the initial key actors have been identified, the analyst can begin to establish the links to other actors and annotate the nature of those links, that is, for example, if the link is a legislatively required referral, or perhaps it is simply a consultative relationship.

Step 4: Finalize and Review ActorMap Diagram

The ActorMap should be constructed whilst the analyst steps through these stages. At this stage, the analyst should review the ActorMap and ensure that all links have been identified and that all annotated links are appropriate. SMEs should be asked to review the ActorMap to ensure its validity. This review and revise stage of the ActorMap process normally requires several iterations.

10.4.4 Advantages

- ActorMaps enable the identification of system-wide actors and their relationships.
- The method is simple to learn and use.
- It is based upon a sound theoretical model.
- Its output offers an exhaustive analysis of actors in systems.
- It provides a clear visual interpretation of the actor relationships, interdependencies, and gaps.
- It is a generic approach that can be applied across many domains.

10.4.5 Disadvantages

- The method can be time-consuming.
- The quality of the analysis produced is entirely dependent upon the quality of the analyst data and investigation.
- ActorMaps simply show the stakeholder relationships.
- Its graphical output can become extensive and hard to decipher when considering complex systems.

10.4.6 Related Methods

To establish thorough ActorMaps, it is important that data collection methods such as interviews and document reviews must be utilized first. ActorMaps analysis is an expanded component of Accimap analysis (Rasmussen, 1997; Svedung and Rasmussen, 2002).

10.4.7 Approximate Training and Application Times

ActorMaps are a simple method to learn and apply and can be used quickly as a sketching and scoping exercise, but they can also become time-consuming when applied to large complex systems. If they are exhaustive, they could take similar timeframes to Accimaps (3 weeks) regardless time should be taken with the final SME verification and validation stage.

10.4.8 Reliability and Validity

There are no reliability and validity data available, but the reliability of the approach may be rigorous if all actors within the systems and their relationships are identified.

10.4.9 Tools Needed

Pen and paper are all that are required for ActorMaps.

10.4.10 Example

ActorMaps have been established in a number of domains as an integral part of the Accimap process. More recently, in a LUP & UD context, they have been used to identify the range of actors associated with the establishment of neighborhoods that support independent living for adults with intellectual disability (MacMillan and Stevens, in press).

The example detailed in the following is that of a study exploring the array of actors in a complex system associated with transport planning. The study revealed an array of 42 actors responsible for the safety and management of beach driving on world heritage listed K'gari (Fraser Island) (Stevens and Salmon, 2015a). Previous reporting of the responsible actors associated with a number of driving-related incidents on the island had only identified 12 actors (Figure 10.14).

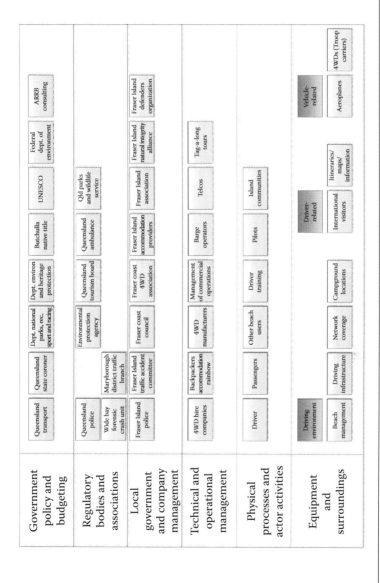

Figure 10.14 ActorMap of beach driving actors. (From Stevens, N. J. and Salmon, P. M., *Proc. Manuf.*, 3, 2605–2612, 2015.)

10.4.11 Recommended Text(s)

Rasmussen, J. (1997). Risk management in a dynamic society: A modelling
    problem. *Safety Science, 27*(2/3), 183–213.
Svedung, J. & Rasmussen, J. (2002). Graphic representation of accident sce-
    narios: Mapping system structure and the causation of accidents. *Safety
    Science, 40*, 397–417.

11

FUTURE HUMAN FACTORS APPLICATIONS IN LAND USE PLANNING AND URBAN DESIGN

11.1 Introduction

Both human factors (HF) and land use planning and urban design (LUP & UD) are at an exciting juncture in terms of their increasing role in the design of our future world. A world that will be increasingly technology centric and developing in areas such as automation, artificial intelligence, and big data. Although the possibilities for optimizing systems and human behavior are tantalizing, the potential to create systems that are cumbersome, inefficient, and even killers is omnipresent.

HF will play a key role in ensuring that future systems do not frustrate, injure, and kill people on a regular basis, whereas LUP & UD has a critical role in the design of the environments in which our future lives will play out. Integrating the two disciplines provides a compelling opportunity to ensure that the design of our future living environments is optimized for human use based on long-standing theory, method, and principles. As such, the possibilities are endless.

The current chapter will highlight the ongoing implications and application of HF methods in LUP & UD. It first considers the past and the ongoing efforts of LUP & UD to elucidate and provide *meaning* to the complexity of urban and regional development. These are practices that have often not resulted in substantive change but instead have provided reflective and normative descriptions of how we have and *should* be living, working, and playing. Next, readers will be challenged to consider the future implications of cities and urban form that prioritize empirically understood models of the complex nature of humans, technology, and their environment. The synergies between the HF and LUP & UD disciplines are strong and we give thought to the role of HF practitioners may play in supporting LUP & UD. The appetite for HF in the LUP & UD domain is clear and the possibilities for knowledge exchange considerable. Further, we contemplate how HF can enhance the wider exchange of knowledge between a range of existing (and necessary) interdisciplinary approaches that collaborate and interface with LUP & UD. To that end, an entirely new paradigm, that of *Sociotechnical Urbanism*, will be introduced. This presents itself as a vital research and practice agenda for future HF and LUP & UD applications.

11.2 Back to the Future

It is possible to identify a long legacy of theoretical, conceptual, and practical efforts that seek to create better cities and interpret their complexity. There have been theoretical paradigms ranging from Walter Christaller's "central place theory" (Christaller, 1966) (proposed in 1933) to Lindblom's (1959) exploration of incrementalism and the science of muddling through. The land use planning discipline has conceived participatory and transactive approaches that draw upon and, indeed, prioritize community input and dialog (Forester, 1999; Friedman, 1973). There are communicative and collaborative approaches that seek to incorporate the range of interests and their associated experiences, including individual subjective participation within urban settings (Healey, 1997). Although the literature around LUP & UD provides insights into how these approaches can be studied and delivered, it has continued to struggle with the identification of practical means or proactive strategies for necessary change.

In the past, there have been a variety of conceptual and applied approaches that have sought to deliver utopian and best-practice models of urban integration. These range from Ebenezer Howard's (1902) Garden City Movement to the enduring *concentric zone model* of city form offered by Ernest Burgess in 1925 or the *multiple nuclei model* of Harris and Ullman (1945). A noteworthy aspect of nearly all utopian models of urban integration is the extent to which they have delivered distinctly dystopian outcomes, more often than not because people in these cities did not behave as expected. Perhaps the classic example is Le Corbusier's *Quartiers Modernes Fruges*, an ideological and esthetic experiment in *prisme pur* or *machines for living*. Far from the expected rationality, Le Corbusier's modernist homes, which were provided for workers of the nearby Fruges factory, were quite quickly personalized, adapted, hacked, or otherwise adjusted to suit the needs and whims of their occupants. This behavior is an inherent property of all sociotechnical systems and a recurring theme in the perceived downfall of most techno-utopian urban planning dreams.

More recently, there has been a range of planning and design *movements* that seek to provide leadership and guidance on good city structures around distinctly human constructs such as quality of life, infrastructure efficiencies, and city futures. Congress for the New Urbanism (1993), for example, advocates quality of life in urban neighborhoods in its charter, foregrounding the mix of different land uses in walkable communities. Transit-orientated developments seek to have commercial, residential, and community facilities, all within the *walkable* proximity of major public transport nodes such as railway stations (Cervero et al., 2002). In the twenty-first century, we recognize and conceive creative (Scott, 2006), smart (Caragliu et al., 2011), or knowledgeable (Ergazakis et al., 2004) cities that focus on the use of human resources, social capital, education, innovation, and information communication and digital technologies. Common to all of them, however, are the often simplified assumptions about how the decisions made by designers and planners will enhance or constrain human behavior within the environment that is subsequently built. We argue that the underlying weakness with many of these approaches is that they tend to provide sets of standards, guidelines, and normative principles for the way-built environments and urban form *should* be designed and developed. Because of this, they often deliver generic and illustrative guidance on best-practice examples, detailing instances from different contexts in the expectation that the ideas may *transplant* to other regions. This translocation of a successful place is often limited in success, as the underlying emergent factors that made it work in one setting are not understood and indeed not transferred to a new location.

All this reflects a key issue for the LUP & UD discipline as a whole. When considering the methodological guidance available to LUP & UD, it is clear there is little literature that assembles practical approaches to data analysis and application. In other words, a strongly empirical and evidence-based approach, particularly with regard to human behavior in built environments, is difficult for the typical practitioner to access. Of course, there are books that deal with urban planning methods in the context of research and policy analysis (Bracken, 2014), whereas broader land use planning texts have largely provided insights into the development and changes to theory and practice over time (Fainstein and Campbell, 2002; Paris, 2013). Alternately, LUP & UD books often represent edited chapters of reflective practice that reiterate the need for planning principles while contemplating potential future city circumstance (Freestone, 2000; Hall and Tewdwr-Jones, 2010). There is some guidance on urban-planning analyses that takes subdisciplinary views such as economics, the environment, or public health. These texts also often outline key principles and normative approaches, and they are largely restricted to a dialog between disciplines and are scant on practical methods of analyses or applications (Brooks et al., 2012; Frumkin et al., 2004).

What has been lacking, until now, for LUP & UD is a suite of accessible methods and the means for all practitioners, researchers and students to explore the inherent complexity of cities. Such complexity will only increase with the advent of *smart* cities, and it is a complexity that stems from an inviolable social and technical interconnection. Past experience shows that we have failed to establish ways to identify the formative possibilities of what *could* happen within our cities. Rather, we continue to be surprised when people in our urban landscapes behave in ways we have clearly *enabled* yet not predicted or considered when designing them. This book is a direct response to this, offering a range of HF methods that may be used in a variety of geographic and jurisdictional contexts. We offer ways for LUP & UD disciplines to begin to explore the use of the voluminous data that our urban communities generate but which we struggle to utilize effectively for sound evidence-based decision-making.

11.3 Beyond Business as Usual

Without appropriate interdisciplinary understanding, LUP & UD may only be expected to generate the inefficient, resource hungry, and placeless urban settings to which we have become accustomed. The ways in which we have designed and delivered our urban and regional environments over the past century cannot continue unchecked. After thousands of years of progress in urban development and organization, we have plateaued. Cities are not safer, healthier, more efficient, or more equitable despite sometimes brave and visionary approaches to changing the status quo. Indeed, they are getting worse. The statistics on fear of crime and chronic disease in our urban environments paints a bleak future as do rising road tolls and congestion and emissions levels. Globally, the accommodation and support infrastructures (i.e., transport, social, and economic) cannot be built fast enough to support the urbanization of the world's population. There is an urgency to understand and plan for the ways in which we want to live in urban communities. We need to explore the cities we have and retrofit and redesign them to accommodate much higher density and more resource-efficient living. Yet, they must still offer the amenity and urban design quality to ensure that they are not simply shelters, machines for living, but homes and communities where life's necessities and services are met. This is not the utopian dream, this is the reality of the situation that must be addressed, and the best way to explore these complex human-centered urban systems is with complex systems methods.

We all agree that urban development and LUP & UD are complex systems. If we consider two aspects of this complexity that have held back LUP & UD, it is possible to identify that HF methods and theoretical approaches may offer the required insights. First, planning must deal with a range of significant issues at a variety of scales, from the site to the local to the regional. There are the human-scale interactions—seating, fittings, and fixtures—right through to significant large-scale exogenous shocks—disasters, technology, or terrorism. There are also a range of complex issues such as transport, water, and even the economy that are crosscutting from the micro to the macro scale, as they traverse formal and informal regulatory and spatial systems. HF methods have the capacity to enhance the understanding of this complexity—that is, both as discrete issues across different spatial and temporal dimensions as well as the interdependence of a range of issues at a particular scale or hierarchy.

Second, LUP & UD involve an increasing range of stakeholders and finance from the public and private sectors. LUP & UD are big businesses and expensive pursuits, requiring commitment for results that are often long term (beyond the political cycles) and frequently in the face of competing public and private sector priorities. Further, we face issues that require collaborative and multidisciplinary approaches, often at different times and often at the same time. Here too, HF can help the consideration of complex and conflicting stakeholder systems. They permit ways for decision-makers to understand the value and evidence in commitment. All participants in the system can be afforded the opportunity to see their places in the puzzle and, indeed, that their priorities are reflected, considered, and optimized. The use of HF methods reveals ways to recognize the consequences of choices and trade-offs and be assisted in understanding the interdependencies and possibilities of better coordination and cooperation in the delivery and design of city systems.

11.4 New Work for Human Factors and Land Use Planning and Urban Design

The message for LUP & UP professionals is clear: HF offers a new approach for achieving a more balanced sociotechnical outcome for current and future cities—a way to capture the human requirements of urban form. The message for HF practitioners and researchers is that there is a deep well of unresolved complex urban issues that can benefit from your attention. These subsume the entirety of day-to-day issues with which you are familiar: whether it is the daily commute, the neighborhood we live in, the parks and open spaces we visit, the redevelopment of the high street, or the urban sprawl that consumes your country. These are the challenges in which your expertise can assist. In being a complex sociotechnical system, the built environment is not only amenable to HF methods, but HF methods, in their turn, offer the possibility to discover small, clever, and inexpensive LUP & UP solutions that yield disproportionately large effects.

Moreover, now is the time to become involved in a new frontier of HF research and application.

For those in parallel disciplines who continue to despair at the state of our cities, and its impacts on human health, wealth, and happiness, HF methods offer a way to explore the synergies and interactions between known complementary fields. The nexus between LUP & UD and transportation systems (active, passive, and autonomous), between LUP & UD and preventative and public health (physical and cognitively engaging cities), between LUP & UD and technology and the arts (our smart and creative cities), the balance between LUP & UD, ecological systems, and the economy—it does not have to be one or the other. We contend that HF methods and HF systems methods particularly offer new ways to explore the possibilities for productive, efficient, and community-centered cities and towns through evidence-based and scientific approaches to LUP & UD.

11.5 A New Agenda—Sociotechnical Urbanism

In a rapidly changing world where smart cities are desired and urban megacities are increasingly a reality, we need to explore the potential for *Sociotechnical Urbanism*. It is a paradigm that builds on the legacy of recognition of the city as system and reinforces the notion that they are indeed sociotechnical systems. The application of HF methods and importantly multiple methods to complex sociotechnical systems provides this new agenda for land use planning and urban design research. It is an agenda to support HF, ergonomics, and the array of LUP & UD researchers and practitioners in exploring, learning, and applying human factors and sociotechnical systems approaches. It is an agenda to allow for the demonstration, tailoring, and detailing of experiences in the practical design and application of these methods for optimizing our urban and regional environments.

Sociotechnical Urbanism, operationalized through the use of HF & STS methods, calls for all stakeholders to begin to conceive and design LUP & UD processes and outcomes that explore and leverage from the recognized properties of city complexity—nonergodicity (lack of probable behavior over time), phase transition (tipping points in the system), emergence (new systems arise from interactions), and universality (despite difference, it is recognized as the same system).

This is the potential for new designs, policy, and processes that recognize complexity and make the constraints we face explicit. It is an approach that is valid for the design and understanding of

- Smaller urban settings—a footpath, a bus stop, parks, and squares
- The neighborhood, the mixed-use urban corridor, and subregional catchments
- Beyond to the city, the state, and global systems and urban challenges like terrorism and climate change

Significantly, it is also the critical and enhanced exploration (at each of these scales) of change over time and the important interim uses of our cities. Understanding and optimizing the inevitable and important short, medium, and long-term transitional phases of urban and regional development.

This is not a flight of fancy, it is necessity, and *Sociotechnical Urbanism* offers a coherent, accessible and comprehensive means to allow for the valid analyses of the urban and regional network of systems and subsystems. The time is right for effective and specific guidance that will assist in addressing some of the perennial and universal challenges of built environment design and analyses. What is certain is that the integration of HF & STS methods in the urban and regional development problem space provides new opportunities for our current and future communities. Constraints, complexity, and emergent behaviors are not necessarily concepts that are ordinarily associated with urban development. However, it is the acknowledgment, identification, and optimization of these systems characteristics that have the most to offer our LUP & UD futures.

References

Adams, M. J., Tenney, Y. J., & Pew, R. W. (1995). Situation awareness and the cognitive management of complex systems. *Human Factors*, *37*(1), 85–104.

Ahlstrom, U. (2005). Work domain analysis for air traffic controller weather displays. *Journal of Safety Research*, *36*(2), 159–169.

Ainsworth, L., & Marshall, E. (1998). Issues of quality and practicality in task analysis: Preliminary results from two surveys. *Ergonomics*, *41*(11), 1604–1617.

Allison, C. K., Revell, K. M., Sears, R., & Stanton, N. A. (2017). Systems theoretic accident model and process (STAMP) safety modelling applied to an aircraft rapid decompression event. *Safety Science*, *98*, 159–166.

Almeida, I. M., & Johnson, C. W. (2005). Extending the borders of accident investigation: applying novel analysis techniques to the loss of the Brazilian space programme's launch vehicle VLS-1 V03. Available at: http://www.dcs.gla.ac.uk/~johnson/papers/Ildeberto_and_Chris.PDF.

Annett, J. (2002). A note on the validity and reliability of ergonomics methods. *Theoretical Issues in Ergonomics Science*, *3*(2), 229–232.

Annett, J. (2004). Hierarchical task analysis. In N. A. Stanton, A. Hedge, K. Brookhuis, E. Salas, & H. Hendrick (Eds.), *Handbook of human factors and ergonomics methods* (pp. 33-1–33-7). Boca Raton, FL: CRC Press.

Annett, J., Duncan, K. D., Stammers, R. B., & Gray, M. J. (1971). Task analysis. *Department of employment training information paper 6*. London: HMSO.

Artman, H., & Garbis, C. (1998). Situation awareness as distributed cognition. In *Proceeding of the 9th European Conference on Cognitive Ergonomics (ECCE'98)*, August (pp. 24–28).

Baber, C. (2005). Evaluating Human-Computer Interaction. In J. R. Wilson & E. N. Corlett (Eds.), *Evaluation of human work* (3rd edn.). Boca Raton, FL: CRC Press.

Baber, C., & Mellor, B. (2001). Using critical path analysis to model multimodal human–computer interaction. *International Journal of Human-Computer Studies*, 54(4), 613–636.

Baber, C., & Stanton, N. A. (1996a). Human error identification techniques applied to public technology: Predictions compared with observed use. *Applied Ergonomics*, 27(2), 119–131.

Baber, C., & Stanton, N. A. (1996b). Observation as a technique for usability evaluation. In P. W. Jordan, B. Thomas, B. A. Weerdmeester, & I. McClelland (Eds.), *Usability evaluation in industry* (pp. 85–94). London: Taylor & Francis Group.

Bainbridge, L. (1995). Verbal protocol analysis. In J. R. Wilson & E. N. Corlett (Eds.) *Evaluation of human work: A practical ergonomics methodology* (pp. 161–179). London: Taylor & Francis Group.

Batty, M. (1971). Modelling cities as dynamic systems. *Nature*, 231(5303), 425–428.

Batty, M. (1976). Models, methods and rationality in urban and regional planning: Developments since 1960. *Area*, 8(2), 93–97.

Batty, M. (2005). *Cities and complexity: Understanding cities with cellular automata, agent-based models, and fractals.* Cambridge, MA: The MIT Press.

Batty, M. (2007). Complexity in city systems: Understanding, evolution, and design. UCL Working Paper Series. UCL Centre for Advanced Spatial Analysis, Paper 117, March.

Batty, M. (2008). The size, scale, and shape of cities. *Science*, 319(5864), 769–771.

Batty, M. (2009). Cities as complex systems: Scaling, interaction, networks, dynamics and urban morphologies. In R.A. Meyers (Ed.), *Encyclopedia of complexity and systems science* (pp. 1041–1071). New York, NY: Springer.

Batty, M. (2013). *The new science of cities.* Cambridge, MA: MIT Press.

Batty, M. (2015). Models again: Their role in planning and prediction. *Environment and Planning B*, 42(2), 191–194.

Billings, C. E. (1995). Situation awareness measurement and analysis: A commentary. *Proceedings of the international conference on experimental analysis and measurement of situation awareness.* Daytona Beach, FL: Embry-Riddle Aeronautical University Press.

Bisantz, A. M., Roth, E., Brickman, B., Gosbee, L. L., Hettinger, L., & McKinney, J. (2003). Integrating cognitive analyses in a large-scale system design process. *International Journal of Human-Computer Studies*, 58(2), 177–206.

Blandford, A., & Wong, B. L. W. (2004). Situation awareness in emergency medical dispatch. *International Journal of Human-Computer Studies*, 61(4), 421–452.

Bødker, S., Ehn, P., Knudsen, J., Kyng, M., & Madsen, K. (1988, January). Computer support for cooperative design. In *Proceedings of the 1988 ACM conference on computer-supported cooperative work* (pp. 377–394). New York, NY: ACM.

Bolstad, C. A., Riley, J. M., Jones, D. G., & Endsley, M. R. (2002). Using goal directed task analysis with Army brigade officer teams. In *Proceedings of the 46th annual meeting of the human factors and ergonomics society* (pp. 472–476). Baltimore, MD.

Braband, J., Evers, B., & de Stefano, E. (2003). Towards a hybrid approach for incident root cause analysis. In *Proceedings of the 21st international system safety conference,* Unionville, VA.

Bracken, I. (2014). *Urban planning methods: Research and policy analysis.* New York, NY: Routledge.

Brooks, N., Donaghy, K., & Knaap, G. J. (2012). *The Oxford handbook of urban economics and planning.* Oxford: Oxford University Press.

Caragliu, A., Del Bo, C., & Nijkamp, P. (2011). Smart cities in Europe. *Journal of Urban Technology, 18*(2), 65–82.

Casali, J. G., & Wierwille, W. W. (1983). A comparison of rating scale, secondary-task, physiological, and primary-task workload estimation techniques in a simulated flight task emphasizing communications load. *Human Factors, 25*(6), 623–641.

Cervero, R., Ferrell, C., & Murphy, S. (2002). *Transit-oriented development and joint development in the United States: A literature review* (p. 52). Washington, DC: TCRP Research Results Digest.

Chin, M., Sanderson, P., & Watson, M. (1999). Cognitive work analysis of the command and control work domain. In *Proceedings of the 1999 command and control research and technology symposium* (Vol. 1, pp. 233–248). Newport, RI.

Christaller, W. (1966). *Central places in southern Germany.* Englewood Cliffs, NJ: Prentice-Hall.

City of Los Angeles. (2008). Walkability Checklist Guidance for Entitlement Review, City of Los Angeles Department of City Planning. http://planning.lacity.org/urbandesign/resources/docs/LAWalkabilityChecklist/lo/LAWalkabilityChecklist-CH01.pdf (accessed 12/10/16).

Corbusier, L. (1967). *The radiant city: Elements of a doctrine of urbanism to be used as the basis of our machine-age civilization.* New York, NY: Orion Press.

Cornelissen, M., Salmon, P. M., Jenkins, D. P., & Lenne, M. G. (2013). A structured approach to the Strategies Analysis phase of Cognitive Work Analysis. *Theoretical Issues in Ergonomics Science, 14*(6), 546–564.

Cornelissen, M., Salmon, P. M., McClure, R., & Stanton, N. A. (2015). Assessing the "system" in safe systems-based road designs: Using cognitive work analysis to evaluate intersection designs. *Accident Analysis and Prevention, 74,* 324–338.

Crandall, B., Klein, G., & Hoffman, R. (2006). *Working minds: A practitioner's guide to cognitive task analysis.* Cambridge, MA: MIT Press.

Cronholm, S. (2009, November). The usability of usability guidelines: A proposal for meta-guidelines. In *Proceedings of the 21st annual conference of the australian computer-human interaction special interest group: Design: Open 24/7* (pp. 233–240). New York, NY: ACM.

Dallat, C., Salmon, P. M., & Goode, N. (2017). Risky systems versus Risky people: To what extent do risk assessment methods consider the systems approach to accident causation? A review of the literature. *Safety Science.*

Dallat, C., Salmon, P. M., & Goode, N. (2017). Identifying risks and emergent risks across sociotechnical systems: the NETworked hazard analysis and risk management system (NET-HARMS). *Theoretical Issues in Ergonomics Science*, 1–27.

Diaper, D. (1989). *Task Analysis in Human Computer Interaction*. Chichester: Ellis Horwood.

Diaper, D., & Stanton, N. S. (2004). *The handbook of task analysis for human-computer interaction*. Mahwah, NJ: Lawrence Erlbaum Associates.

Djokic, J., Lorenz, B., & Fricke, H. (2010). Air traffic control complexity as workload driver. *Transportation Research Part C: Emerging Technologies*, *18*(6), 930–936.

Dominguez, C. (1994). Can SA be defined? In M. Vidulich, C. Dominguez, E. Vogel, & G. McMillan (Eds.), *Situation Awareness: Papers and annotated bibliography*. Report AL/CF-TR-1994-0085. Wright-Patterson Airforce Base, OH: Air Force Systems Command.

Donovan, S. L., Salmon, P. M., Lenné, M. G., & Horberry, T. (2017). Safety leadership and systems thinking: Application and evaluation of a risk management framework in the mining industry. *Ergonomics*, 1–15.

Driskell, J. E., & Mullen, B. (2004). Social network analysis. In N. A. Stanton, A. Hedge, K. Brookhuis, E. Salas, & H. Hendrick (Eds.), *Handbook of human factors and ergonomics methods* (pp. 58.1–58.6). Boca Raton, FL: CRC Press.

Drury, C. (1990). Methods for direct observation of performance. In J. R. Wilson & E. N. Corlett (Eds.), *Evaluation of human work*, (pp. 45–68). London: Taylor & Francis Group.

Durlauf, S. N. (2005). Complexity and empirical economics. *The Economic Journal*, *115*(504), 225–243.

Durso, F. T., Hackworth, C. A., Truitt, T., Crutchfield, J., & Manning, C. A. (1998). Situation awareness as a predictor of performance in en route air traffic controllers. *Air Traffic Quarterly*, *6*(1), 1–20.

Easterby, R. (1984). Tasks, processes and display design. In R. Easterby & H. Zwaga (Eds.), *Information design* (pp. 19–36). Chichester, West Sussex: Whiley.

Embrey, D. E. (1986). SHERPA: A systematic human error reduction and prediction approach. *Paper presented at the International Meeting on Advances in Nuclear Power Systems*, Knoxville, TN.

Endsley, M. R. (1993). A survey of situation awareness requirements in air-to-air combat fighters. *The International Journal of Aviation Psychology*, *3*(2), 157–168.

Endsley, M. R. (1995a). Towards a theory of situation awareness in dynamic systems. *Human Factors*, *37*(1), 32–64.

Endsley, M. R. (1995b). Measurement of situation awareness in dynamic systems. *Human Factors*, *37*(1), 65–84.

Endsley, M. R. (2016). *Designing for situation awareness: An approach to user-centered design*. Boca Raton, FL: CRC press.

Endsley, M. R., & Jones, W. M. (1997). Situation awareness, information dominance, and information warfare. Technical Report 97-01. Belmont, MA: Endsley Consulting.

Endsley, M. R., & Robertson, M. M. (2000). Situation awareness in aircraft maintenance teams. *International Journal of Industrial Ergonomics*, *26*(2), 301–325.

Endsley, M. R., Sollenberger, R., & Stein, E. (2000). Situation awareness: A comparison of measures. In *Proceedings of the human performance, situation awareness and automation: User-centered design for the new millennium*. Savannah, GA: SA Technologies.

Ergazakis, K., Metaxiotis, K., & Psarras, J. (2004). Towards knowledge cities: Conceptual analysis and success stories. *Journal of Knowledge Management*, *8*(5), 5–15.

Erlandsson, M., & Jansson, A. (2007). Collegial verbalisation—A case study on a new method on information acquisition. *Behaviour & Information Technology*, *26*(6), 535–543.

Eysenck, M. W., & M. T. Keane. (1990). *Cognitive psychology: A student's handbook*. Hove: Lawrence Erlbaum.

Fainstein, S. S., & Campbell, S. (Eds.). (2002). *Readings in urban theory*. Oxford: Blackwell.

Faludi, A. (1973). *Planning theory*. Oxford: Pergamon Press.

Federal Aviation Administration, FAA. (1996) *Report on the Interfaces between Flightcrews and Modern Flight Deck Systems,* Federal Aviation Administration, Washington DC, 1996.

Finstad, K. (2010). The usability metric for user experience. *Interacting with Computers*, *22*(5), 323–327.

Flanagan, J. C. (1954). The critical incident technique. *Psychological Bulletin*, *51*(4), 327–358.

Forester, J. (1999). *The deliberative practitioner: Encouraging participatory planning processes*. Cambridge, MA: MIT Press.

Fracker, M. (1991). *Measures of situation awareness: Review and future directions* (Rep. No. AL-TR-1991-0128). Wright Patterson Air Force Base, OH: Armstrong Laboratories, Crew Systems Directorate.

Freestone, R. (2000). *Urban planning in a changing world: The twentieth century experience*. New York, NY: Taylor & Francis Group.

Friedman, J. (1973). *Retracking America: A theory of transactive planning*. Garden City, NJ: Anchor Press/Doubleday.

Frumkin, H., Frank, L., & Jackson, R. J. (2004). *Urban sprawl and public health: Designing, planning, and building for healthy communities*. Washington, DC: Island Press.

Gorman, J. C., Cooke, N., & Winner, J. L. (2006). Measuring team situation awareness in decentralised command and control environments. *Ergonomics*, *49*(12–13), 1312–1326.

Gray, W. D., John, B. E., & Atwood, M. E. (1993). Project ernestine: Validating a GOMS analysis for predicting and explaining real-world task performance. *Human-Computer Interaction, 8*(3), 237–309.

Hall, P., & Tewdwr-Jones, M. (2010). *Urban and regional planning.* London: Routledge.

Harris, C. D., & Ullman, E. L. (1945). The nature of cities. *The Annals of the American Academy of Political and Social Science, 242*(1), 7–17.

Harris, D., Stanton, N. A., Marshall, A., Young, M. S., Demagalski, J., & Salmon, P. M. (2005). Using SHERPA to predict design induced error on the flight deck. *Aerospace Science and Technology, 9*(6), 525–532.

Hart, S. G., & Staveland, L. E. (1988). Development of a multi-dimensional workload rating scale: Results of empirical and theoretical research. In P. A. Hancock & N. Meshkati (Eds.), *Human mental workload.* Amsterdam: Elsevier.

Harvey, C., & Stanton, N. A. (2013). *Usability evaluation for in-vehicle systems.* Boca Raton, FL: CRC Press.

Hauss, Y., & Eyferth, K. (2003). Securing future ATM-concepts' safety by measuring situation awareness in air traffic control. *Aerospace Science and Technology, 7*(6), 417–427.

Hazlehurst, B., McMullen, C. K., & Gorman, P. N. (2007). Distributed cognition in the heart room: How situation awareness arises from coordinated communications during cardiac surgery. *Journal of Biomedical Informatics, 40*(5), 539–551.

Healey, P. (1997). *Collaborative planning: Shaping places in fragmented societies.* Vancouver, Canada: UBC Press.

Heinrich, H. W. (1931). *Industrial Accident Prevention.* New York: McGraw-Hill.

Hoffman, K. A., Aitken, L. M., & Duffield, C. (2009). A comparison of novice and expert nurses' cue collection during clinical decision-making: Verbal protocol analysis. *International Journal of Nursing Studies, 46*(10), 1335–1344.

Hogg, D. N., Folleso, K., Strand-Volden, F., & Torralba, B. (1995). Development of a situation awareness measure to evaluate advanced alarm systems in nuclear power plant control rooms. *Ergonomics, 38*(11), 2394–2413.

Holling, C. S., & Goldberg, M. A., (1971). Ecology and planning. *Journal of the American Institute of Planners, 37*(4), 221–230.

Hollnagel, E. (1998). *Cognitive reliability and error analysis method—CREAM,* (1st ed.). Oxford, England: Elsevier Science.

Hollnagel, E. (2012). *FRAM: The functional resonance analysis method: Modelling complex socio-technical systems.* Aldershot: Ashgate.

Hopkins, A. (2005). *Safety, Culture and Risk: The Organisational Causes of Disasters.* Sydney, CC H.

Houghton, R. J., Baber, C., Cowton, M., Stanton, N. A., & Walker, G. H. (2008). WESTT (Workload, error, situational awareness, time and teamwork): An analytical prototyping system for command and control. *Cognition Technology and Work, 10*(3), 199–207.

Houghton, R. J., Baber, C., McMaster, R. et al. (2006). Command and control in emergency services operations: A social network analysis. *Ergonomics*, *49*(12–13), 1204–1225.

Howard, E. (1902). *Garden cities of tomorrow*. London: S. Sonnenschein & Co., Ltd.

Hughes, C. M., Baber, C., Bienkiewicz, M., Worthington, A., Hazell, A., & Hermsdörfer, J. (2015). The application of SHERPA (Systematic human error reduction and prediction approach) in the development of compensatory cognitive rehabilitation strategies for stroke patients with left and right brain damage. *Ergonomics*, *58*(1), 75–95.

Hutchins, E. (1995). *Cognition in the wild*. Cambridge, MA: MIT Press.

Isaac, A., Shorrock, S. T., Kennedy, R., Kirwan, B., Anderson, H., & Bove, T. (2002). Technical review of human performance models and taxonomies of error in air traffic management (HERA). Eurocontrol Project Report, HRS/HSP-002-REP-01.

Jacob, S. A., & Furgerson, S. P. (2012). Writing interview protocols and conducting interviews: Tips for students new to the field of qualitative research. *The Qualitative Report*, *17*(42), 1–10.

James, N., & Patrick, J. (2004). The role of situation awareness in sport. In S. Banbury & S. Tremblay (Eds.), *A cognitive approach to situation awareness: Theory and application* (pp. 297–316), Aldershot: Ashgate Publishing Limited.

Jansson, A., Olsson, E., & Erlandsson, M. (2006). Bridging the gap between analysis and design: Improving existing driver interfaces with tools from the framework of cognitive work analysis. *Cognition, Technology & Work*, *8*(1), 41–49.

Jenkins, D. P. (2009). *Cognitive work analysis: Coping with complexity*. Farnham: Ashgate Publishing.

Jenkins, D. P., Salmon, P. M., Stanton, N. A., & Walker, G. H. (2010). A systemic approach to accident analysis: a case study of the Stockwell shooting. *Ergonomics* 53(1), 1–17.

Jenkins, D. P., Stanton, N. A., Walker, G. H., & Salmon, P. M. (2008). *Cognitive work analysis: Coping with complexity*. Aldershot: Ashgate.

Karsh, B. T., Waterson, P., & Holden, R. J. (2014). Crossing levels in systems ergonomics: A framework to support "mesoergonomic" inquiry. *Applied Ergonomics*, *45*(1), 45–54.

Karwowski, W. (2001). *International Encyclopedia of Ergonomics and Human Factors Vols I–III*. London: Taylor & Francis Group.

Kilgore, R., & St-Cyr, O. (2006). The SRK inventory: A tool for structuring and capturing a worker competencies analysis. *Human factors and ergonomics society annual meeting proceedings, cognitive engineering and decision making* (pp. 506–509). San Francisco, CA: HFES.

Kirwan, B., & Ainsworth, L. K. (1992). *A guide to task analysis*. London: Taylor & Francis Group.

Kirwan, B., Evans, A., Donohoe, L. et al. (1997). *Human factors in the ATM system design life cycle* (pp. 16–20). Paris, France: FAA/Eurocontrol ATM R&D Seminar, June.

Klein, G., & Armstrong, A. A. (2004). Critical decision method. In N. A. Stanton, A. Hedge, E. Salas, H. Hendrick, & K. Brookhaus (Eds.), *Handbook of human factors and ergonomics methods* (pp. 35.1–35.8). Boca Raton, FL: CRC Press.

Klein, G., Calderwood, R., & McGregor, D. (1989). Critical decision method for eliciting knowledge. *IEEE Transactions on Systems, Man & Cybernetics, 19*(3), 462–472.

Kleiner, B. M. (2006). Macroergonomics: Analysis and design of work systems. *Applied Ergonomics, 37*(1), 81–89.

Langford, J., & McDonagh, D. (2002). Focus Groups: Supporting Effective Product Development. London: Taylor & Francis Group.

Lawton, R., & Ward, N. J. (2005). A systems analysis of the Ladbroke Grove rail crash. *Accident Analysis and Prevention* 37, 235–44.

Lee, S., Moh, Y. B., Tabibzadeh, M., & Meshkati, N. (2017). Applying the AcciMap methodology to investigate the tragic Sewol Ferry accident in South Korea. *Applied ergonomics, 59,* 517–525.

Leveson, N. (2004). A new accident model for engineering safer systems. *Safety Science, 42*(4), 237–270.

Leveson, N. G. (2002). *System safety engineering: Back to the future.* Cambridge, MA: Massachusetts Institute of Technology.

Li, W. C., Harris, D., Hsu, Y.L., & Li, L. W. (2009). The application of the Human Error Template (HET) for redesigning standard operating procedures in aviation operations. In D. Harris (Ed.), *Engineering psychology and cognitive ergonomics.* Oklahoma: Springer Verlag, pp. 600–605.

Lindblom, C. E. (1959). The science of "muddling through."*Public Administration Review, 19*(2), 79–88.

Lockyer, K. G., & Gordon, J. (1991). *Critical path analysis and other project network techniques.* New York: Beekman Books Incorporated.

Lund, A. M. (2001). Measuring usability with the USE questionnaire12. *Usability Interface, 8*(2), 3–6.

Lynch, K. (1960). *The image of the city* (Vol. 11). Cambridge, MA: MIT press.

Malekpour, F., Mohammadian, Y., Malekpour, A. R., Mohammadpour, Y., Sheikh Ahmadi, A., & Shakarami, A. (2014). Assessment of mental workload in nursing by using NASA-TLX. *Journal of Urmia Nursing and Midwifery Faculty, 11*(11), 892–899.

Marshall, A., Stanton, N., Young, M., Salmon, P., Harris, D., Demagalski, J., Waldmann, T., & Dekker, S. (2003). Development of the human error template – a new methodology for assessing design induced errors on aircraft flight decks. ERROR PRED Final Report E!1970, August.

Matthews, M. D., Strater, L. D., & Endsley, M. R. (2004). Situation awareness requirements for infantry platoon leaders. *Military Psychology, 16,* 149–161.

McGuinness, B., & Foy, L. (2000). A subjective measure of SA: The crew awareness rating scale (CARS). *Presented at the human performance, situational awareness and automation conference.* Savannah, GA, October 16–19.

McIlroy, R., C., & Stanton, N. A. (2011). Observing the observer: non-intrusive verbalisations using the Concurrent Observer Narrative Technique. *Cognition, Technology and Work, 13,* 135–149.

McLoughlin, B. (1969). *Urban and regional planning: A systems approach.* London: Faber and Faber.

Megaw, T. (2005). The definition and measurement of mental workload. In J. R. Wilson & N. Corlett (Eds.), *Evaluation of human work*, (pp. 525–553), Boca Raton, FL: CRC Press.

Militello, L. G., & Hutton, J. B. (2000). Applied cognitive task analysis (ACTA): A practitioner's toolkit for understanding cognitive task demands. In J. Annett & N. S. Stanton (Eds.), *Task analysis* (pp. 90–113). London: Taylor & Francis Group.

Miller, J. E., Patterson, E. S., & Woods, D. D. (2006). Elicitation by critiquing as a cognitive task analysis methodology. *Cognition, Technology & Work, 8*(2), 90–102.

Miller, R. B. (1953). *A method for man-machine task analysis: WADC technical report 53–137, Wright air development center air research and development command*, Wright-Patterson Air Force Base, OH.

Mills, S. (2007). Contextualising design: Aspects of using usability context analysis and hierarchical task analysis for software design. *Behaviour & Information Technology, 26*(6), 499–506.

Missen, L., Stevens, N. J., & Salmon, P. M. (2017). A sociotechnical systems analysis approach to playground design. In *Contemporary ergonomics 2017, proceedings of the chartered institute of ergonomics and human factors annual conference*, April, Daventry.

Mulvihill, C. M., Salmon, P. M., Beanland, V., Lenné, M. G., Read, G. J., Walker, G. H., & Stanton, N. A. (2016). Using the decision ladder to understand road user decision making at actively controlled rail level crossings. *Applied Ergonomics, 56*, 1–10.

Naikar, N., Hopcroft, R., & Moylan, A. (2005). *Work domain analysis: Theoretical concepts and methodology.* Defence Science & Technology Organisation Report, DSTO-TR-1665, Melbourne, Australia.

Naikar, N., Moylan, A., & Pearce, B. (2006). Analysing activity in complex systems with cognitive work analysis: Concepts, guidelines, and case study for control task analysis. *Theoretical Issues in Ergonomics Science, 7*(4), 371–394.

Naikar, N., Pearce, B., Drumm, D., & Sanderson, P. (2003a). Designing teams for first-of-a-kind, complex systems using the initial phases of cognitive work analysis: case study. *Human Factors* 45(2), 202–17.

Naikar, N., & Sanderson, P. M. (1999). Work domain analysis for training-system definition. *International Journal of Aviation Psychology, 9*(3), 271–290.

Naikar, N., & Sanderson, P. M. (2001). Evaluating design proposals for complex systems with work domain analysis. *Human Factors, 43*(4), 529–542.

Naikar, N., & Saunders, A. (2003b). Crossing the boundaries of safe operation: A technical training approach to error management. *Cognition Technology and Work, 5*, 171–180.

Neisser, U. (1976). *Cognition and reality: Principles and implications of cognitive psychology.* San Francisco, CA: Freeman.

Neville, T. J., & Salmon, P. M. (2016). Never blame the umpire–A review of situation awareness models and methods for examining the performance of officials in sport. *Ergonomics*, *59*(7), 962–975.

Noyes, J. M. (2006). Verbal protocol analysis. In W. Karwowski (Ed.), *International encyclopaedia of ergonomics and human factors* (2nd ed.), (pp. 3390–3392). London: Taylor & Francis Group.

Nygren, T. E. (1991). Psychometric properties of subjective workload measurement techniques: Implications for their use in the assessment of perceived mental workload. *Human Factors*, *33*(1), 17–33.

O'Hare, D., Wiggins, M., Williams, A., & Wong, W. (2000). Cognitive task analysis for decision centred design and training. In J. Annett & N. A. Stanton (Eds.), *Task analysis* (pp. 170–190). London: Taylor & Francis Group.

Omodei, M. M., & McLennan, J. (1994). Studying complex decision making in natural settings: Using a headmounted video camera to study competitive orienteering. *Perceptual and Motor Skills*, 79(3), 1411–1125.

Oppenheim, A. N. (2000). *Questionnaire design, interviewing and attitude measurement*. London: Continuum.

Ottino, J. M. (2003). Complex systems. *AIChE Journal*, *49*(2), 292–299.

Paris, C. (Ed.). (2013). *Critical readings in planning theory: Urban and regional planning series*. Amsterdam, the Netherlands: Elsevier.

Patorniti, N. P., Stevens, N. J., & Salmon, P. M. (2017). A systems approach to city design: Exploring the compatibility of sociotechnical systems. *Habitat International*, 66, 42–48.

Patrick, J., Gregov, A., & Halliday, P. (2000). Analysing and training task analysis. *Instructional Science*, *28*(4), 57–79.

Patrick, J., James, N., Ahmed, A., & Halliday, P. (2006). Observational assessment of situation awareness, team differences and training implications. *Ergonomics*, *49*(12–13), 393–417.

Peute, L. W., & Jaspers, M. W. (2007). The significance of a usability evaluation of an emerging laboratory order entry system. *International Journal of Medical Informatics*, *76*(2), 157–168.

Phipps, D., Meakin, G. H., Beatyy, P. C. W., Nsoedo, C., & Parker, D. (2008) Human Factors in anaesthesia practice: Insights from a task analysis. *British Journal of Anaesthesia* 100(3), 333–43.

Polson, P. G., Lewis, C., Rieman, J., & Wharton, C. (1992). Cognitive walkthroughs: A method for theory-based evaluation of user interfaces. *International Journal of Man-Machine Studies*, *36*(5), 741–773.

Pratt, J. (2017). *The real walkable catchment: What elements affect walkability amongst different age groups? Unpublished honours theses. Bachelor of regional and urban planning* (*Honours*), Sunshine Coast, Queensland: University of the Sunshine Coast.

Qureshi, Z. H. (2007, December). A review of accident modelling approaches for complex socio-technical systems. In *Proceedings of the twelfth Australian workshop on safety critical systems and software and safety-related programmable systems* (Vol. 86, pp. 47–59). Australian Computer Society.

Rasmussen, J. (1985). The role of hierarchical knowledge representation in decision making and system management. *IEEE Transactions on Systems, Man and Cybernetics, 15,* 234–243.

Rasmussen, J. (1997). Risk management in a dynamic society: A modelling problem. *Safety Science, 27*(2/3), 183–213.

Rasmussen, J., Pejtersen, A. M., & Goodstein, L. P. (1994). *Cognitive systems engineering.* New York, NY: Wiley.

Ravden, S. J., & Johnson, G. I. (1989). *Evaluating usability of human-computer interfaces: A practical method.* Chirchester, West Sussex: Ellis Horwood.

Read, G. J. M., Beanland, V., Lenné, M. G., Stanton, N. A., & Salmon, P. M. (2017). *Integrating human factors methods and systems thinking for transport analysis and design.* Boca Raton, FL: Taylor & Francis Group.

Read, G. J. M., Salmon, P. M., & Lenné, M. G. (2016). When paradigms collide at the road rail interface: Evaluation of a sociotechnical systems theory design toolkit for cognitive work analysis. *Ergonomics, 59*(9), 1135–1157.

Read, G., Salmon, P. M., Lenne, M. G., & Jenkins, D. P. (2015). Designing a ticket to ride with the cognitive work analysis design toolkit. *Ergonomics, 58*(8), 1266–1286.

Read, J. M. G., Salmon, P. M., & Lenne, M. G. (2013). Sounding the warning bells: The need for a systems approach to understanding behaviour at rail level crossings. *Applied Ergonomics, 44,* 764–774.

Reason, J. (1990). *Human error.* New York, NY: Cambridge University Press.

Reason, J. (1997). *Managing the risks of organisational accidents.* Burlington, VT: Ashgate Publishing.

Reid, G. B., & Nygren, T. E. (1988). The subjective workload assessment technique: A scaling procedure for measuring mental workload. In P. S. Hancock, & N. Meshkati (Eds.), *Human mental workload.* Amsterdam, the Netherlands: Elsevier.

Robie, C., Brown, D. J., & Beaty, J. C. (2007). Do people fake on personality inventories? A verbal protocol analysis. *Journal of Business and Psychology, 21*(4), 489–509.

Roscoe, A., & Ellis, G. (1990). *A subjective rating scale for assessing pilot workload in flight.* Farnborough, Hampshire: RAE.

Rowley, J. (2012). Conducting research interviews. *Management Research Review, 35*(3/4), 260–271.

Royal Australian Aviation Force. (2001). *F111 Deseal/ Reseal Board of Inquiry Report.* Available at: http://www.airforce.gov.au/docs/Volume1.htm.

Salas, E., Prince, C., Baker, P. D., & Shrestha, L. (1995). Situation awareness in team performance. *Human Factors, 37,* 123–136.

Salmon, P. M., Dallat, C., & Clacy, A. (2017a). More than just the bike: Distributed situation awareness and teamwork in elite women's cycling teams. In *Contemporary ergonomics 2017, proceedings of the chartered institute of ergonomics and human factors annual conference*, April, Daventry.

Salmon, P. M., Goode, N., Archer, F. L., Spencer, C., McArdle, D., & McClure, R. J. (2014b). A systems approach to understanding and enhancing disaster response: An Accimap analysis of bushfire response activities. *Safety Science*, *70*, 114–122.

Salmon, P. M., Jenkins, D., Stanton, N., & Walker, G. (2010). Hierarchical task analysis versus cognitive work analysis: Comparison of theory, methodology and contribution to system design. *Theoretical Issues in Ergonomics Science*, *11*(6), 504–531.

Salmon, P. M., Lenné, M. G., Read, G. J., Mulvihill, C. M., Cornelissen, M., Walker, G. H., & Stanton, N. A. (2016b). More than meets the eye: Using cognitive work analysis to identify design requirements for future rail level crossing systems. *Applied Ergonomics*, *53*, 312–322.

Salmon, P. M., Lenné, M. G., Walker, G. H., Stanton, N. A., and Filtness, A., (2014a). Exploring schema-driven differences in situation awareness between road users: An on-road study of driver, cyclist and motorcyclist situation awareness. *Ergonomics*, *57*(2), 191–209.

Salmon, P. M., Read, G. J., Lenne, M., Mulvihill, C., Stevens, N. J., Walker, G. H., Young, K., & Stanton, N. A., (2017c). Cognitive work analysis for systems analysis and redesign: Rail level crossing case study. In *Cognitive work analysis: Applications, extensions and the future*. Boca Raton, FL: CRC Press.

Salmon, P. M., Read, G. J., & Stevens, N. J. (2016a). Who is in control of road safety? A STAMP control structure analysis of the road transport system in Queensland, Australia. *Accident Analysis & Prevention*, *96*, 140–151.

Salmon, P. M., Stanton, N. A., Jenkins, D. P., & Walker, G. H. (2011). Coordination during multi-agency emergency response: Issues and solutions. *Disaster Prevention and Management*, *20*(2), 140–158.

Salmon, P. M., Stanton, N. A., Walker, G. H., Baber, C., Jenkins, D. P., & McMaster, R. (2008a). What really is going on? Review of situation awareness models for individuals and teams. *Theoretical Issues in Ergonomics Science*, *9*(4), 297–323.

Salmon, P. M., Stanton, N. A., Walker, G. H., & Green, D. (2006). Situation awareness measurement: A review of applicability for C4i environments. *Applied Ergonomics*, *37*(2), 225–238.

Salmon, P. M., Stanton, N. A., Walker, G. H., & Jenkins, D. P. (2009). *Distributed situation awareness: Advances in theory, measurement and application to teamwork*. Aldershot: Ashgate.

Salmon, P. M., Stanton, N. A., Walker, G. H., Jenkins, D. P., Baber, C., & McMaster, R. (2008b). Representing situation awareness in collaborative systems: A case study in the energy distribution domain. *Ergonomics*, *51*(3), 367–384.

Salmon, P. M., Stanton, N. A., Young, M. S., Harris, D., Demagalski, J., Marshall, A., Waldman, T., & Dekker, S. (2002) Using existing HEI techniques to predict pilot error: a comparison of SHERPA, HAZOP and HEIST. In *Proceedings of HCI-Aero 2002*. Cambridge, MA : MIT Press.

Salmon, P. M., Walker, G. H., Read, G. J. M., Goode, N., & Stanton, N. A. (2017b). Fitting methods to paradigms: Are ergonomics methods fit for systems thinking? *Ergonomics*, *60*(2), 194–205.

Sarter, N. B., & Woods, D. D. (1991). Situation awareness—A critical but ill-defined phenomenon. *International Journal of Aviation Psychology*, *1*(1), 45–57.

Scott, A. J. (2006). Creative cities: Conceptual issues and policy questions. *Journal of Urban Affairs*, *28*(1), 1–17.

Shadbolt, N. R., & Burton, M. (1995). Knowledge elicitation: A systemic approach. In J. R. Wilson, & E. N. Corlett (Eds.), *Evaluation of human work: A practical ergonomics methodology*, (pp. 406–440). London: Taylor & Francis Group.

Shepherd, A. (2002) *Hierarchical task analysis*. London: Taylor & Francis Group.

Shorrock, S. T., & Kirwan, B. (2002). Development and application of a human error identification tool for air traffic control. *Applied Ergonomics*, *33*, 319–336.

Shu, Y., & Furuta, K. (2005). An inference method of team situation awareness based on mutual awareness. *Cognition Technology & Work*, *7*(4), 272–287.

Siemieniuch, C. E., & Sinclair, M. A. (2006). Systems integration. *Applied Ergonomics*, *37*(1), 91–110.

Smith, K., & Hancock, P. A. (1995). Situation awareness is adaptive, externally directed consciousness. *Human Factors*, *37*(1), 137–148.

Sonnenwald, D. H., Maglaughlin, K. L., & Whitton, M. C. (2004). Designing to support situation awareness across distances: An example from a scientific collaboratory. *Information Processing and Management*, *40*(6), 989–1011.

Stanton, N. A. (2005). Hierarchical Task Analysis: developments, extensions and applications. *Applied Ergonomics* 37(1), 55–79.

Stanton, N. A. (2005b). Human Factors and ergonomics methods. In N. A. Stanton, A. Hedge, K. Brookhuis, E. Salas, & H. Hendrick (Eds.), *Handbook of human factors and ergonomics methods*. Boca Raton, FL: CRC Press.

Stanton, N. A. (2006). Hierarchical task analysis: Developments, applications, and extensions. *Applied Ergonomics*, *37*(1), 55–79.

Stanton, N. A. (2014). Representing distributed cognition in complex systems: How a submarine returns to periscope depth. *Ergonomics*, *57*(3), 403–418.

Stanton, N. A., & Baber, C. (1996). A systems approach to human error identification. *Safety Science*, *22*, 215–28.

Stanton, N. A., & Baber, C. (2002). Error by design: Methods for predicting device usability. *Design Studies*, *23*(4), 363–384.

Stanton, N. A., Baber, C., & Harris, D. (2008). *Modelling command and control: Event analysis of systemic teamwork*. Aldershot: Ashgate.

Stanton, N. A., & Bessell, K. (2014). How a submarine returns to periscope depth: Analysing complex socio-technical systems using cognitive work analysis. *Applied Ergonomics*, *45*(1), 110–125.

Stanton, N. A., Harris, D., Salmon, P. M. et al. (2006b). Predicting design induced pilot error using HET (Human Error Template)—A new formal human error identification method for flight decks. *Journal of Aeronautical Sciences*, *110*(1104), 107–115.

Stanton, N. A., & Harvey, C. (2017). Beyond human error taxonomies in assessment of risk in sociotechnical systems: A new paradigm with the EAST "broken-links" approach. *Ergonomics*, *60*(2), 221–233.

Stanton, N. A., Hedge, A., Brookhuis, K., Salas, E., & Hendrick, H. (2004). *Handbook of human factors methods*. Boca Raton, FL: CRC Press.

Stanton, N. A., Jenkins, D. P., Salmon, P. M., Walker, G. H., Revell, K., & Rafferty, L. A. (2016). *Digitising command and control: A human factors and ergonomics analysis of mission planning and battlespace management*. Boca Raton, FL: CRC Press.

Stanton, N. A., & McIlroy, R. C. (2012). Designing mission communication planning: the role of Rich Pictures and cognitive work analysis. *Theoretical Issues in Ergonomics Science*, *13*(2), 146–168.

Stanton, N. A., McIlroy, R. C., Harvey, C., Blainey, S., Hickford, A., Preston, J. M., & Ryan, B. (2013b). Following the cognitive work analysis train of thought: Exploring the constraints of modal shift to rail transport. *Ergonomics*, *56*(3), 522–540.

Stanton, N. A., Salmon, P. M., Rafferty, L., Walker, G. H., Jenkins, D. P., & Baber, C. (2013a). *Human factors methods: A practical guide for engineering and design* (2nd ed.). Aldershot: Ashgate.

Stanton, N. A., Salmon, P. M., & Walker, G. H. (2015). Let the reader decide: A paradigm shift for situation awareness in sociotechnical systems. *Journal of Cognitive Engineering and Decision Making*, *9*(1), 44–50.

Stanton, N. A., Salmon, P. M., Walker, G., Baber, C., & Jenkins, D. P. (2005). *Human factors methods: A practical guide for engineering and design*. Aldershot: Ashgate.

Stanton, N. A., Salmon, P. M., Walker, G. H., Hancock, P. A., & Salas, E. (2017b). State-of-science: Situation awareness in individuals, teams and systems. *Ergonomics*, *60*(4), 449–466.

Stanton, N. A., Salmon, P. M., Walker, G. H., & Jenkins, D. P. (2009). Genotype and phenotype schemata and their role in distributed situation awareness in collaborative systems. *Theoretical Issues in Ergonomics Science*, *10*(1), 43–68.

Stanton, N. A., Salmon, P. M., Walker, G. H., & Jenkins, D. P. (2017a). *Cognitive work analysis: Applications, extensions and future*. Boca Raton, FL: CRC Press.

Stanton, N. A., & Stevenage, S. V. (1998). Learning to predict human error: Issues of acceptability, reliability and validity. *Ergonomics*, *41*(11), 1737–1747.

Stanton, N. A., Stewart, R., Harris, D. et al. (2006a). Distributed situation awareness in dynamic systems: Theoretical development and application of an ergonomics methodology. *Ergonomics*, *49*, 1288–1311.

Stanton, N. A., & Young, M. S. (1998). Is utility in the mind of the beholder? A review of ergonomics methods. *Applied Ergonomics*, *29*(1), 41–54.

Stanton, N. A., & Young, M. (1999). *A guide to methodology in ergonomics: Designing for human use*. London: Taylor & Francis Group.

Stanton, N. A., & Young, M. (1999a). *A guide to methodology in ergonomics: Designing for human use*. London: Taylor & Francis Group.

Stanton, N. A., & Young, M. S. (1999b). What price ergonomics? *Nature*, *399*, 197–198.

Stanton, N. A., & Young, M. S. (2003). Giving ergonomics away? The application of ergonomics methods by novices. *Applied Ergonomics*, *34*(5), 479–490.

Stanton, N. A., Young, M. S., & Harvey, C. (2014). *Guide to methodology in ergonomics: Designing for human use*. Boca Raton, FL: CRC Press.

Stevens, N. J. (2016). Sociotechnical urbanism: New systems ergonomics perspectives on land use planning and urban design. *Theoretical Issues in Ergonomics Science*, *17*(4), 443–451.

Stevens, N., & Salmon, P. (2014). Safe places for pedestrians: Using cognitive work analysis to consider the relationships between the engineering and urban design of footpaths. *Accident Analysis & Prevention*, *72*, 257–266.

Stevens, N. J., & Salmon, P. M. (2015a). All responsibility, no care: A systems analysis case study of beach driving stakeholders in Australia. *Procedia Manufacturing*, *3*, 2605–2612.

Stevens, N., & Salmon, P. (2015b). New knowledge for built environments: Exploring urban design from socio-technical system perspectives. In *Engineering psychology and cognitive ergonomics* (pp. 200–211). New York, NY: Springer International Publishing.

Stevens, N. J., & Salmon, P. M. (2016). Sand, sun and sideways: A systems analysis of beach driving. *Safety Science*, *85*, 152–162.

Stevens, N. J., Salmon, P. M., & Taylor, N. (2017). Work domain analysis applications in urban planning: Active transport infrastructure and urban corridors. In *Cognitive work analysis: Applications, extensions and the future*. Boca Raton, FL: Taylor & Francis Group.

Stevens, N., Salmon, P., Walker, G., & Stanton, N. (2016, September). Off the beaten track: Situation awareness in experienced and novice off-road drivers. In *Proceedings of the 2016 australasian road safety conference*. Australasian College of Road Safety.

Stewart, R., Stanton, N. A., Harris, D. et al. (2008). Distributed situation awareness in an airborne warning and control system: Application of novel ergonomics methodology. *Cognition Technology and Work*, *10*(3), 221–229.

Stoner, H. A., Wiese, E. E., & Lee, J. D. (2003, October). Applying ecological interface design to the driving domain: The results of an abstraction hierarchy analysis. In *Proceedings of the human factors and ergonomics society annual meeting*, (Vol. 47, No. 3, pp. 444–448). Sage CA: Los Angeles, CA: SAGE Publications.

Svedung, I., & Rasmussen, J. (2002). Graphic representation of accident scenarios: Mapping system structure and the causation of accidents. *Safety Science*, *40*(5), 397–417.

Taylor, R. M. (1990). Situational awareness rating technique (SART): The development of a tool for aircrew systems design. In *Situational awareness in aerospace operations (AGARD-CP-478)*, (pp. 3/1–3/17). Neuilly Sur Seine, France: NATO-AGARD.

Uhlarik, J., & Comerford, D. A. (2002). *A review of situation awareness literature relevant to pilot surveillance functions*. (DOT/FAA/AM-02/3). Washington, DC: Federal Aviation Administration, U.S. Department of Transportation.

UN (United Nations). (2014). World urbanization prospects. *Department of economic and social affairs, population division. The 2014 revision, highlights (ST/ESA/SER.A/352)*. New York, NY: United Nations.

Vicente, K. J. (1999). *Cognitive work analysis: Toward safe, productive, and healthy computer-based work*. Mahwah, NJ: Lawrence Erlbaum Associates.

Vidulich, M. A., & Hughes, E. R. (1991). Testing a subjective metric of situation awareness. *In Proceedings of the human factors society 35th annual meeting* (pp. 1307–1311). Santa Monica, CA: Human Factors Society.

Vidulich, M. A., & Tsang, P. S. (1985). *Collecting NASA workload ratings*. Moffett Field, CA: NASA Ames Research Center.

Vidulich, M. A., & Tsang, P. S. (1986). Technique of subjective workload assessment: A comparison of SWAT and the NASA bipolar method. *Ergonomics*, *29*(11), 1385–1398.

Waag, W. L., & Houck, M. R. (1994). Tools for assessing situational awareness in an operational fighter environment. *Aviation, Space and Environmental Medicine*, *65*(5), A13–A19.

Waldrop, M. M. (2015). No drivers required. *Nature*, *518*(7537), 20.

Walker, G. H. (2004). Verbal protocol analysis. In N. A. Stanton et al., (Eds.), *The handbook of human factors and ergonomics methods*. Boca Raton, FL: CRC Press.

Walker, G. H., Gibson, H., Stanton, N. A., Baber, C., Salmon, P. M., & Green, D. (2006). Event analysis of systemic teamwork (EAST): A novel integration of ergonomics methods to analyse C4i activity. *Ergonomics*, *49*, 1345–1369.

Walker, G. H., Stanton, N. A., Baber, C., Wells, L., Jenkins, D. P., & Salmon, P. M. (2010). From ethnography to the EAST method: A tractable approach for representing distributed cognition in air traffic control. *Ergonomics*, *53*(2), 184–197.

Walker, G. H., Salmon, P. M., Bedinger, M., & Stanton, N. A. (2017). Quantum ergonomics: Shifting the paradigm of the systems agenda. *Ergonomics*, *60*(2), 157–166.

Walker, G. H., & Stanton, N. A. (2017). *Human factors in automotive engineering and technology.* London: CRC Press.

Walker, G. H., Stanton, N. A., & Young, M. S. (2001). Hierarchical task analysis of driving: A new research tool. In M. A. Hanson (Ed.), *Contemporary ergonomics* (pp. 435–440). London: Taylor & Francis Group.

White, E. (1983). *Site Analysis: Diagramming Information for Architectural Design,* Architectural Media, Florida.

Wickens, C. D. (1992). *Engineering psychology and human performance* (2nd ed.). New York, NY: Harper Collins.

Wiegmann, D. A., & Shappell, S. A. (2003). *A human error approach to aviation accident analysis: The human factors analysis and classification system.* Burlington, VT: Ashgate Publishing.

Wierwille, W. W., & Eggemeier, F. T. (1993). Recommendations for mental workload measurement in a test and evaluation environment. *Human Factors, 35*(2), 263–282.

Wilson, J. R., & Corlett, N. (1995). *Evaluation of human work—A practical ergonomics methodology* (2nd ed.). London: Taylor & Francis Group.

Wong, T. H. (2006). Water sensitive urban design-the journey thus far. *Australian Journal of Water Resources, 10*(3), 213–222.

Young, M. S., Brookhuis, K. A., Wickens, C. D., & Hancock, P. A. (2015). State of science: Mental workload in ergonomics. *Ergonomics, 58*(1), 1–17.

Index

Printed and bound by CPI Group (UK) Ltd, Croydon, CR0 4YY

17/10/2024

01775682-0017